P9-DUZ-012

STRUCTURAL THERMODYNAMICS OF ALLOYS

JACK MANENC

STRUCTURAL THERMODYNAMICS OF ALLOYS

D. REIDEL PUBLISHING COMPANY

DORDRECHT-HOLLAND / BOSTON-U.S.A.

THERMODYNAMIQUE STRUCTURALE DES ALLIAGES

First published by Presses Universitaires de France, Paris, 1972

Translated from the French by N. Corcoran

Library of Congress Catalog Card Number 73-83564

ISBN 90 277 0346 9

Published by D. Reidel Publishing Company,
P.O. Box 17, Dordrecht, Holland

Sold and distributed in the U.S.A., Canada, and Mexico
by D. Reidel Publishing Company, Inc.
306 Dartmouth Street, Boston,
Mass. 02116, U.S.A.

All Rights Reserved
Copyright © 1973 by D. Reidel Publishing Company,
Dordrecht, Holland
No part of this book may be reproduced in any form, by print,
photoprint, microfilm, or any other means,
without written permission from the publisher

Printed in The Netherlands by D. Reidel, Dordrecht

LIBRARY
University of Texas
At San Antonio

WITHDRAWN
UTSA LIBRARIES

CONTENTS

PART II: STRUCTURE OF ALLOYS NOT IN THE EQUILIBRIUM STATE

INTRODUCTION

Technical progress has for a very long time been directly dependent on progress in metallurgy, which is itself connected with improvements in the technology of alloys. Metals are most frequently used in the form of alloys for several reasons: the quantity of pure metal in its native state in the earth's crust is very limited; pure metals must be extracted from ores which are themselves impure. Finally, the methods of treatment used lead more easily to alloys than to pure metals. The most typical case is that of iron, where a pure ore may be found, but which is the starting point for cast iron or steel, alloys of iron and carbon.

In addition, the properties of alloys are in general superior to those of pure metals and modern metallurgy consists of controlling these properties so as to make them conform to the requirements of the design office.

Whilst the engineer was formerly compelled to adapt his designs and constructions to the materials available, such as wood, stone, bronze, iron, cast iron and ordinary steels, he can now expect, due to metallurgical research, the creation of special alloys meeting specific requirements. These requirements must of course be reasonable, but must be sufficiently imperative for them to become the motive for progress.

Thus, starting from cast iron and steels, manufacture of which had been improved so as to give better control over the mechanical properties, we reached the industrial purification of iron, which led to the development of alloys with properties very much superior to those of ordinary steels, thus opening up very great future prospects.

This book will not deal with elaboration, although this is a subject of very great technical importance. Nor will we consider electronic structure, since this does not appear essential for the object that we have set ourselves. This object is a description of the structural aspect, since, as we shall see, it is by acting on the microstructure that we determine the mechanical or physical properties. The electronic theory of metals should however enable us in the near future to predict these properties, starting from the properties of the constituents. It does not at present appear to us to be sufficiently advanced to be a great deal of use.

We have divided the book into three parts. A few pages after the introduction will deal with definitions. One part will treat the thermodynamic aspect of constitution in equilibrium. Another part will be devoted to the phase transformations leading to the characteristic structures.

The thermodynamic aspect of the structure of alloys must, as for all physical phenomena, be taken into account to reach an understanding of the behaviour observed or to draw up new details or new thermal treatments in accordance with the principles of this branch of physics. It forms in effect a framework containing the physical phenomena. No macroscopic changes can take place in conflict with its principles and its theorems. It is in some manner a guide which indicates to the engineer and to the physicist what is possible.

A knowledge of thermodynamics brings even more. It enables us to predict metastable phenomena, of which the interest is apparent in Part II. If we had this complete knowledge, we would be able to choose the fine details and the compositions without the time-consuming experiments which are still necessary.

The evolution of the microstructure is a subject of the greatest importance. Man has in fact discovered empirically that at high temperatures the properties of alloys change with temperature, the change being more rapid at high temperatures. It is by suitable choice of temperature, heating time, speed of heating or cooling, the number and the sequence of heat treatments that he has learned to invent new mechanical characteristics, the great number of which makes it possible always to select one which meets the needs of any technical construction.

If the first metallurgists were thus able to provide their alloys with exceptional properties, then modern metallurgists are beginning to understand scientifically how such results are obtained.

GENERAL CONSIDERATIONS

1. Constituents and Phases

In Part I we will consider the laws which govern the mixing of metals. These latter will then be regarded as the constituents of the alloy. This term 'constituent' usually has a wider meaning and can refer to definite compounds. Here, it will mean the pure elements which constitute the alloy.

We will also use the term 'phase'. *A phase is a part of a system which is chemically and physically homogeneous, and is separated from the other parts by an interface.*

A homogeneous liquid forms one phase; a solid may consist of one or several phases. Two non-miscible liquids form a two-phase system (oil and water, for example).

Vapours and gases, in the pure state or as mixtures, form only one phase.

2. Definitions of an Alloy

A crystalline substance composed of two or more elements having the characteristics of the metallic state is known as a metallic alloy. Metallic alloys, just as metals, differ in many respects from other crystals. One of their particular properties is electrical conductivity linked to free movement of valency electrons. A metal consists in fact of ionised atoms, of which the positive charge is neutralised by the presence of electrons which do not belong to any of the atoms in particular.

As far as the thermodynamic properties are concerned, at least for a first approximation, we do not concern ourselves with the atoms and we neglect the effect of the free electrons. The electrons contribute little to the specific heat of metals and alloys.

Alloys may consist of any number of elements, but current theories deal effectively only with binary or pseudo-binary alloys.

Alloys of metals and metalloids are considered to be metallic alloys when the percentage of metalloid is small.

To produce an alloy, it is necessary to put the constituents together and make them react. If there is no reaction, we have a mechanical mixture. When the reaction has taken place and the alloy is produced, we obtain a mixture of phases closely intertwined in direct contact along an interface, whilst with a mechanical mixture the crystals or particles are separated by vacuum or by air.

Alloys in the liquid state are most frequently made up of one single phase. The atoms of the constituents are then distributed at random throughout the volume or they may have a slight tendency to repel or to attract each other.

3. Structure of Alloys in the Solid State

From the crystallographic point of view, solid alloys may be classified in two categories: solid solutions and intermetallic compounds.

There are two types of solid solution: solid solutions of substitution and solid solutions of insertion.

Solid solutions of substitution derive their structure from the base metal from which they are formed. As for metals, metallic solid solutions are body-centred cubic, b.c.c., face-centred cubic, f.c.c., hexagonal close packed, h.c.p., and more rarely tetragonal, t.

In the b.c.c. structures, the atoms are distributed at random at the apices and at the centre of an elementary cube of side a few Ångström units in length, which constitutes the unit of repetition.

In the f.c.c. structure, the atoms are distributed at random at the apices and at the centres of the faces of the cube forming the unit of repetition.

The hexagonal structure consists of atoms forming a compact arrangement on a base plane.

Lastly, the tetragonal structure comprises atoms placed at random at the apices of a rectangular parallelopiped with a square base.

The structures of insertion-type solid solutions are also

derived from those for pure metals. But the atoms of the solute no longer replace the atoms of the solvent in the positions in the lattice; they are situated in the spaces left between the atoms.

This can take place only for small atoms, such as carbon, nitrogen, hydrogen, oxygen and boron. The most instructive case is that of iron where carbon enters as an insertion type solution.

Thus at high temperatures, face-centred cubic iron can dissolve a large quantity of carbon – about 1.8% by weight at 1100° C.

On the other hand, for body-centred cubic iron at low temperatures, 0.002% by weight of carbon can enter into solution in equilibrium.

Although the face-centred cubic structure is more compact than the body-centred cubic, it contains larger interstices than this latter. For the face-centred cubic lattice these interstices are at the centre of the cube and at the centre of the tetrahedra formed by four neighbouring atoms. For the body-centred cubic lattice they are on the edges and at the centre of the tetrahedra formed by four neighbouring atoms.

4. Intermetallic or Definite Compounds

The structure of certain phases of alloys is uniquely derived from the metallic structures by ordered arrangement of the atoms in position. This leads to a chemical formula which is usually simple, of type M_xN_y, where x and y are small integers.

For example, in the Al-Cu system we find the compound Al_2Cu, and in the Ni–Al system the phases Ni_3Al and NiAl. However, we find more and more compounds where the formula is not simple. These may then be regarded as intermetallic solid solutions of small extent.

Contrary to the solid solutions mentioned above, the formation of definite compounds has its origin in the large difference in properties of the atoms forming the alloy. These differences may be chemical, electronic or steric in nature. It is difficult to isolate one or other effect, since it is quite certain that the properties of an atom in solution or in a metallic compound are dependent on its electronic structure and on the modifications to that structure resulting from the association.

The bonds in intermetallic compounds may be partially metallic and partially ionic. In this latter case the atoms have different electronegativities.

5. Definition of the Microstructure of Alloys

Like metals and indeed many solid crystals, alloys are made up of triply periodic arrangements of the unit cell, body-centred, face-centred, or hexagonal, to which we have referred above. A unit cell is defined by its crystallographic axes which enable us to determine its orientation. A single crystal consists of a very large number of unit cells of which the position in space is fixed by the vector sums of a whole number of elementary translations parallel to each other. The monocrystalline state is extremely rare amongst metals and alloys. These solids are most frequently found in the polycrystalline state. They are then composed of a very large number of small crystals with their axes not parallel to each other. The crystals forming a polycrystalline sample are closely attached along a surface known as a grain boundary.

A grain boundary is a fault inside one phase. It does not constitute a separation interface between two phases from the thermodynamic point of view. Composition and structure are in fact identical on each side of the boundary; only the orientation changes.

Grain boundaries may be classified in three categories: twin boundaries, small-angle boundaries and large-angle boundaries.

Twin boundaries are those which separate grains where the relative orientation is well defined and symmetrical about the boundary itself. This latter is then plane and parallel to a low-index crystallographic plane, $\{111\}$ for face-centred cubic or $\{112\}$ for body-centred cubic structures. The twin boundary consists of a monatomic layer; the atoms forming it belong to both twinned crystals.

Small-angle boundaries are those where the reticular planes of one family make an angle less than 10° on the two sides of the joint.

Large-angle boundaries are least well understood. Their

structure is probably very complex and cannot be described by simple diagrams. Contrary to what was formerly believed, their thickness is at most a few atomic layers.

It is clear that grain boundaries, especially those where there is a large change in orientation, are places where the structure of the metal is very complicated and where the atoms are in relative positions which are different from those which they occupy within the grain. This fact is of great importance for alloys, since it is at the interfaces between the grains that segregation of added elements most frequently takes place. Precipitation takes place most frequently at these same places, as we shall see during our discussion of hardening of alloys by precipitation. It would theoretically be possible to eliminate grain boundaries completely by making monocrystalline materials. This would have the advantage of eliminating the areas of low mechanical resistance which they produce when covered by the precipitated phase. Unfortunately, it is rarely possible to grow grains of sufficient size.

PART I

EQUILIBRIUM STATES

CHAPTER 1

REVIEW OF FUNDAMENTAL CONCEPTS OF CLASSICAL THERMODYNAMICS OF SOLUTIONS

1.1. Stable and Metastable Equilibria

The concept of thermodynamic equilibrium is fundamental. It applies as much to mechanics as to all physical phenomena. Equilibrium corresponds to the most stable state of a system. Any slight change from this state leads to an absorption of energy which makes the system in an unstable state; it then reverts spontaneously to the initial state by liberating the energy.

We also find metastable states in nature. In this case, evolution towards a more stable state liberates energy but is not spontaneous.

The definition of the stable state is really valid only for an isolated system. If this system is placed in the universe, its state is then unstable. Let us take as an example a rectangular parallelopiped. It is in metastable equilibrium when resting on one of its small faces and in stable equilibrium when resting on a large face. Nevertheless, this solid, placed on a table, is in metastable equilibrium with regard to the earth. It is therefore absolutely necessary to define the limits of the system of which we are considering the equilibrium.

For reasons of convenience we refer to thermal, chemical, electrical and mechanical equilibria, etc., corresponding to the particular aspects of physics, but thermodynamics enables us to treat them all by means of the same general equations.

The equilibrium in which we are interested here is chemical equilibrium. We shall need to concern ourselves above all with the changes in properties resulting from local variations in composition, temperature, pressure and volume of alloys.

1.2. Thermodynamic Functions

Let us consider any system capable of evolution, the state of which depends only on volume, temperature and pressure, surrounded by a medium which we shall call environment (env.) and with which it can interchange energy.

Internal energy

The first law of thermodynamics, which expresses the conservation of energy, tells us that our system will have an internal energy which is the sum of heat energy and potential energy, and which remains constant if no energy is exchanged with the environment. It is an isolated system. The first law tells us also that if an exchange does take place, the change in internal energy of the system is equal to the quantity of heat received less the work done by the system against external forces. We might think by analogy with mechanics (potential energy) that the system is in a state which is more stable the smaller its internal energy. This is not true in general. Depending on the conditions of change, the system can move towards equilibrium by increasing or by decreasing its internal energy.

Suppose we supply an infinitesimal increase of heat $\mathrm{d}Q$ and of work $\mathrm{d}W$ to an isolated system, the energy of the system increases by:

$$\mathrm{d}E = \mathrm{d}Q + \mathrm{d}W\,.$$

This internal energy depends only on the thermodynamic state of the system. The same is true for the *enthalpy function* which is defined by the relationship:

$$\mathrm{d}H = \mathrm{d}E + P\mathrm{d}V = \mathrm{d}Q + \mathrm{d}W + P\mathrm{d}V\,.$$

Entropy

A knowledge of the variations of the functions E and H is not sufficient to define the system from the view-point of equilibrium; on the other hand, a knowledge of the entropy defined by the relationship:

$$\mathrm{d}S = \mathrm{d}Q/T$$

for the *system and its environment*, is sufficient.

This follows from the second law of thermodynamics, which is based on the limitations in conversion of heat energy into mechanical energy. The entropy S is a function of state, which like E and H does not depend on the path taken to reach the state in question.

For an isolated system, the change in this function is either positive or zero, depending on whether the change is irreversible or reversible. If the system is not isolated, the direction of the change depends on interchanges with the surrounding medium. We then show that if the process is reversible, its change is equal but of opposite sign to that of the environment, or if the process is irreversible, it is greater than that of the environment with its sign changed:

$$\Delta S_{syst} = -\Delta S_{env} \quad \text{reversible}$$
$$\Delta S_{syst} > -\Delta S_{env} \quad \text{irreversible}.$$

We can regard theoretically a system and a limited region of its environment as forming an isolated system where the change in entropy can only be positive or zero. We can then say whether a change is possible (it is natural) or impossible (it would be unnatural). It is, however, difficult in practice to calculate or to measure the changes in the entropy of the environment, and we only know the entropy of the system itself, which is not isolated, and this is not sufficient to determine the direction of the change.

It was therefore necessary to seek for other functions of state which could give information more directly about the direction of change. These functions are:

(a) the Gibbs free energy function;
(b) the Helmholtz free energy or free enthalpy.

Gibbs free energy

Consider for example a spontaneous transformation of a system at constant temperature and pressure. We have:

$$\mathrm{d}S_{\mathrm{syst}} + \mathrm{d}S_{\mathrm{env}} \geqslant 0 .$$

Even if the reaction is irreversible, heat is exchanged with the environment medium and we have:

$$\mathrm{d}S_{\mathrm{env}} = -\,\mathrm{d}Q/T ,$$

where $\mathrm{d}Q$ is the heat received by the system.

We also have, under isothermal conditions and at constant pressure:

$$\mathrm{d}H = \mathrm{d}E + P\mathrm{d}V$$
$$\mathrm{d}E = \mathrm{d}Q - P\mathrm{d}V ,$$

whence:

$$\mathrm{d}H = \mathrm{d}Q$$
$$\mathrm{d}S_{\mathrm{env}} = -\,\mathrm{d}H/T \quad \text{and} \quad \mathrm{d}S_{\mathrm{syst}} > \mathrm{d}H/T .$$

Putting:

$$\mathrm{d}H - T\mathrm{d}S_{\mathrm{syst}} < 0$$

for a reversible reaction, we would have:

$$\mathrm{d}H - T\,\mathrm{d}S_{\mathrm{syst}} = 0 ,$$

and for an impossible reaction:

$$\mathrm{d}H - T\,\mathrm{d}S_{\mathrm{syst}} > 0 .$$

The sign of the quantity $\mathrm{d}H—T\,\mathrm{d}S_{\mathrm{syst}}$ can thus serve to show whether a reaction is possible or not in a non-isolated system.

Since we shall in future no longer make use of functions relating to the environment, we shall discontinue the suffixes.

The relation $G=H—TS$ is a function of state, since at constant temperature H and S are themselves functions of state:

$$G = H - TS. \tag{1}$$

G is known as the Gibbs free energy.

Since all natural reactions are possible only if $\mathrm{d}G \leqslant 0$, *stable equilibrium of the system will be reached only when the function G is at a minimum.*

Helmholtz free energy

Another function of state is also defined at constant volume, known as the Helmholtz free energy:

$$F = E - TS. \tag{2}$$

In the same way this function of state must be at a minimum for stable equilibrium at constant volume to be reached.

We now have sufficient relationships to define the state of a system and to predict its spontaneous change, but we must extend the validity of these functions to cases where parameters other than temperature, volume and pressure are concerned. What is of interest to us is the variation in composition. The Gibbs free energy is then expressed as a function of temperature, pressure and concentration of the elements of which the solid or liquid is formed.

Stable equilibrium will be reached whenever G or F are at minima with regard to a change in composition, this being either at constant temperature and pressure or at constant temperature and volume.

CHAPTER 2

VARIATION OF THERMODYNAMIC FUNCTIONS WITH COMPOSITION AT CONSTANT PRESSURE OR VOLUME AND TEMPERATURE

2.1. Partial Molar Functions

When two constituents are mixed to form a solution, the thermodynamic properties are not in general the algebraic sums of the properties of the constituents mechanically mixed.

For example, and this is the simplest case to consider, the volume of the mixture is different from the total volume of the pure constituents before formation of the solution. After mixing a contraction is produced if the forces of attraction between neighbouring atoms are larger in the solution than in the pure constituents, or an expansion in the opposite case.

The extensive thermodynamic functions (volume, internal energy, entropy and free energy) depend in a non-linear manner on the concentrations of the elements. We can however express each thermodynamic function of this type by a linear sum of terms:

$$U(T, P, n_1, \ldots, n_m),$$

and at constant P and T:

$$U(n_i) = \sum \bar{U}_i n_i,$$

where n_i is the number of i gram-atoms in the solution and $\bar{U}_i$ is the corresponding partial molar quantity.

In fact, if we consider a new system in which there are $tn_1, tn_2, \ldots, tn_i, \ldots tn_m$ atoms, we will have a new function $U(\ldots, tn_i, \ldots)$; but since the concentration has not changed, we can put:

$$U(\ldots, tn_i, \ldots) = tU(\ldots, n_i, \ldots).$$

This shows that U is a homogeneous function of the first degree of the n_i's. Let us put:

$$tn_i = u_i,$$

and apply Euler's calculus:

$$U(u_i) = tU(n_i),$$

and differentiating with respect to t, we have:

$$\sum \frac{\partial U}{\partial u_i} \frac{\mathrm{d}u_i}{\mathrm{d}t} = U(n_i),$$

but we also have:

$$\mathrm{d}u_i/\mathrm{d}t = n_i,$$

which gives:

$$\sum \frac{\partial U}{\partial u_i} n_i = U(n_i).$$

Since this expression is valid for all values of t other than zero or infinity, we can put $t=1$, giving:

$$\partial U/\partial u_i = \partial U/\partial n_i,$$

and

$$U(n_i) = \sum \frac{\partial U}{\partial n_i} n_i .$$

For simplicity we agree to write:

$$\partial U/\partial n_i = \bar{U}_i \quad \text{and} \quad U = \sum \bar{U}_i n_i .$$

If we add $\mathrm{d}n_i$ at constant P and T, we find:

$$U + \mathrm{d}U = \sum \bar{U}_i (n_i + \mathrm{d}n_i),$$

whence:

$$\mathrm{d}U = \sum \frac{\partial U}{\partial n_i} \mathrm{d}n_i = \sum \bar{U}_i \, \mathrm{d}n_i .$$

At constant temperature and pressure, the partial differential coefficients $\bar{U}_i$ are homogeneous functions of degree zero with respect to the number of moles of each element. It is for this reason that we prefer to express them by means of molar or atomic fractions, which are also of degree zero with respect to these same variables. In the case where we are concerned with one gram-atom of solution, let:

$$N = \sum n_i \quad \text{and} \quad x_i = n_i / \sum n_i = n_i / N ,$$

and we can write the thermodynamic function per gram-atom:

$$U = \sum \bar{U}_i x_i .$$

The functions $\bar{U}_i$ are magnitudes of quality or of tension, independent of mass as are temperature and pressure.

2.2. Gibbs–Duhem Relationship

Let us differentiate the function U. We obtain:

$$\mathrm{d}U = \sum \mathrm{d}\bar{U}_i n + \sum \bar{U}_i \,\mathrm{d}n_i .$$

We had obtained above:

$$\mathrm{d}U = \sum \bar{U}_i \,\mathrm{d}n_i$$

with the result that

$$\boxed{\sum n_i \mathrm{d}\bar{U}_i = 0 .}$$

This relationship is of great practical importance, in particular for binary alloys. It enables us to calculate the partial molar function of one of the two elements if we know that of the other as a function of the concentration; this is carried out analytically or graphically depending on the particular case. The following method is used:

$$\mathrm{d}\bar{U}_2 = -\frac{n_i}{n_2}\,\mathrm{d}\bar{U}_1 ,$$

whence:

$$\bar{U}_2 - \bar{U}_2^0 = -\int_{\bar{U}_1{}^0}^{\bar{U}_1} \frac{n_i}{n_2}\,\mathrm{d}\bar{U}_1 .$$

Note the practical importance of the expression

$$\mathrm{d}U = \sum \mathrm{d}n_i \bar{U}_i .$$

If we are dealing with a very large quantity of solution, and if we add to this an infinitesimal quantity of one of the elements, the overall composition is not appreciably altered, nor are the values of $\bar{U}_i$ derived from them. This enables us to determine, at effectively constant composition, each of the terms on the right-hand side by a very small change in concentration of each element *i*.

2.3. Expression of the Various Thermodynamic Functions in Terms of the Partial Molar Functions

When the function U denotes the free energy G, the partial derivatives $\bar{G}_i$ are identical with those known as μ_i, chemical potentials, to chemists, and which express the respective tendencies of the constituents to leave the phase in which they are found:

$$G = \sum \mu_i n_i \qquad \bar{G}_i = \mu_i$$

at constant pressure and temperature.

We can express the other functions in the same manner:

volume: $V = \sum \bar{V}_i n_i$
enthalpy: $H = \sum \bar{H}_i n_i$
entropy: $S = \sum \bar{S}_n n_i$
....

and we also have the differential relationships:

$$\sum n_i \,\mathrm{d}\bar{V}_i = 0$$
$$\sum n_i \,\mathrm{d}\bar{H}_i = 0$$
$$\sum n_i \,\mathrm{d}\bar{S}_i = 0 .$$

To conclude, let us recall the relationships between free energy, enthalpy and entropy:

$$S = -\,(\partial G/\partial T)_P \tag{3}$$

$$H = -\,T^2 \partial(G/T)/\partial T = [\partial(G/T)/\partial 1/T]_P \tag{4}$$

$$V = (\partial G/\partial P)_T \tag{5}$$

$$P = -\,(\partial E/\partial V)_T . \tag{6}$$

CHAPTER 3

CLASSIFICATION OF DIFFERENT TYPES OF HOMOGENEOUS SOLUTION ACCORDING TO EXPRESSION OF FREE ENERGY

3.1. General Considerations

As we have seen, the formation of a homogeneous solution results from intimate mixing of the atoms. These atoms are located in space, on the crystalline sites for solids and inside a certain volume for liquids. Their distribution over these positions or at a location in the volume is random, and the probability of occupation of a position or of a location in the volume is, for one type of atom, proportional to its atomic fraction.

The properties of solid solutions are very variable and depend on the properties of their constituents. If the properties of these latter differ only slightly, the properties of the solutions differ little from those of the pure constituents. In the limiting case, the formation of the mixture is accompanied only by insignificant changes in the thermodynamic functions, but the phase resulting from solution of two pure elements is more stable than the mechanical mixture of these elements. The simplest case of solutions of very similar bodies is that of isotopes of the same element.

What takes place in the general case?

Let A and B be two elements capable of forming a solution, a liquid solution or a homogeneous solid. Three cases can arise:

(1) The atoms of the same nature have attractive forces equal to those of the atoms of different nature. From the chemical point of view, the atoms A and B are indistinguishable. The solution is ideal.

(2) Atoms of the same nature attract each other more than atoms of different nature. There is a tendency towards segregation.

(3) Atoms of different nature attract each other more than atoms of the same nature. There is a tendency towards the formation of an order.

3.2. Activity of Elements in Solution

The elements which constitute a solution, liquid or solid, have a vapour pressure which in the pure state is a function of temperature; we will denote it by P_i^0.

When an element contains another element in solution, its vapour pressure is no longer P_i^0 but P_i; P_i is less than P_i^0.

In most cases the vapour may be regarded as a perfect gas, since the *pressures exerted are very small.* Under these conditions, *we define the activity of element i by the ratio*:

$$a_i = P_i/P_i^0 .$$

The value of the activity enables us to give a numerical

measure of the tendency of an element to escape from the phase which contains it.

When the vapours of the constituents are in equilibrium above the solution, the free energy of the whole is at a minimum and we may write:

$$G = \sum \bar{G}_{ig} n_{ig} + \sum \bar{G}_{is} n_{is},$$

where the index g refers to vapours in the gaseous phase and the index s to the solid or liquid phases.

If we cause a change only in concentration in the solid and in the gas of element i, without altering the composition of the whole, we shall have:

$$\delta G = \bar{G}_{ig} \mathrm{d} n_{ig} + \bar{G}_{is} \mathrm{d} n_{is}.$$

For equilibrium, we must have:

$$\delta G = 0.$$

But, since $\mathrm{d}n_{ig} = -\mathrm{d}n_{is}$, we can put for element i:

$$\bar{G}_{is} = \bar{G}_{ig}.$$

If we are dealing with pure constituents, it follows that:

$$\bar{G}_{is} - \bar{G}_{is}^0 = \Delta \bar{G}_{is} = \Delta \bar{G}_{ig} = \bar{G}_{ig} - \bar{G}_{ig}^0 = \Delta \bar{G}_i.$$

The chemical potential of an element in solution is equal in the solution and in the gas above the solution.

Since the saturating vapour may be regarded as a perfect gas we can calculate $\Delta \bar{G}_i$ per mole:

$$(\partial \bar{G}/\partial P)_P = V = RT/P \qquad \mathrm{d}\bar{G} = RT\, \mathrm{d}(\log P),$$

and on integrating:

$$\bar{G}_i - \bar{G}_i^0 = \Delta\bar{G}_i = RT \log(P_i/P_i^0) = \mu_i - \mu_0^0 = \\ = RT \log a_i .$$

3.3. Ideal Solutions. Raoult's Law

In the case of an ideal solution, and this is one of its definitions, the activity is equal to the atomic concentration of the element. In fact, if we express the vapour pressure P_B of the element B as a function of X_B in the ideal solution, we find that it varies linearly from 0 to P_B^0. The activity then varies from 0 to 1.

We therefore have in general:

$$a_i = P_i/P_i^0 = X_i .$$

This relationship is called *Raoult s law*.

We deduce from it in accordance with (4):

$$\frac{\partial \Delta\bar{G}_i/T}{\partial 1/T} = \Delta\bar{H}_i \qquad R\left(\frac{\partial \log X_i}{\partial 1/T}\right) = 0,$$

since X_i does not depend on temperature.

We also deduce from this law the following relationships for one mole of solution:

Enthalpy

$$\Delta H = \sum \Delta\bar{H}_i X_i = 0 .$$

The change in enthalpy on formation of the solution is zero.

In an ideal solid or liquid solution, the component i behaves as if it were in the pure state.

Entropy

Since

$$\Delta G_m = RT \sum X_i \log X_i ,$$

and

$$\Delta G_m = \Delta H - T\Delta S = - T\Delta S$$

at constant temperature and pressure; the change in entropy of the mixture ΔS_m is given by:

$$\boxed{\Delta S_{m(\text{ideal})} = - R \sum X_i \log X_i .}$$

It is found in the same way that $\Delta V_{m\,(\text{ideal})}=0$; *the volume of the ideal solution is equal to the sum of the volumes of the elements before formation of the solution.*

3.4. Non-Ideal Solutions. General Considerations. Dilute Solution. Henry's Law

In general solutions do not follow Raoult's law, which states that the activity is equal to the atomic concentration.

Nevertheless, for a dilute solution, the distance between atoms of the solute is large and interaction between atoms of the same nature is small. The changes in the thermodynamic properties following the introduction of the solute are then proportional to the concentration.

In particular, the vapour pressure of the solute is proportional to the atomic concentration of this latter:

$$\boxed{P_i = \alpha X_i\,,}$$

where α is a constant for the solution. This relationship expresses Henry's law, which is written in terms of the activity:

$$\boxed{a_i = \gamma_i X_i\,,}$$

where γ_i, the activity coefficient of the element i in dilute solution, is constant for very small values of X_i; it may be less than or greater than unity.

When the solution is not very dilute, we can still express the activity by a similar relationship:

$$\boxed{a_i = \gamma_i X_i\,,}$$

but γ_i depends on the concentrations of the elements present in the solution.

How does the solvent behave in dilute solution?

Let the suffix S refer to the solvent and the suffix i refer to one element in dilute solution.

We have seen that the partial molar free energy of the elements follows the Gibbs–Duhem law:

$$X_s \mathrm{d}\bar{G}_S + \sum X_i \,\mathrm{d}\bar{G}_i = 0\,.$$

Now:

$$\Delta\bar{G}_i = RT \log a_i \qquad \mathrm{d}\bar{G}_i = RT\,\mathrm{d}(\log a_i)\,.$$

By substitution:

$$X_s \, \mathrm{d}(\log a_S) + \sum X_i \, \mathrm{d}(\log a_i) = 0 .$$

Since

$$a_i = \gamma_i X_i \qquad (\gamma_i \text{ constant})$$
$$\mathrm{d}(\log a_i) = \mathrm{d}(\log X_i)$$
$$X_s \, \mathrm{d}(\log a_S) = - \sum \mathrm{d}X_i$$
$$\mathrm{d}(\log a_S) = - (1/X_S) \sum \mathrm{d}X_i .$$

Now:

$$X_S + \sum X_i = 1 \qquad \mathrm{d}X_S = - \sum \mathrm{d}X_i$$
$$\mathrm{d}(\log a_S) = \mathrm{d}X_S / X_S$$
$$\mathrm{d}(\log a_S) = \mathrm{d}(\log X_S) .$$

On integrating from 1 to X_S we find, knowing that when $X_S = 1$, $a_S = 1$,

$$\boxed{a_S = X_S .}$$

Thus, *for a dilute solution the solvent obeys Raoult's law.* We can not, however, state a priori the range of values of the composition for which this approximations is valid. This depends very much on the properties of the constituents.

We will now give graphs to illustrate the preceeding for the cases of binary solutions of elements A and B.

In Figure 1 the elements A and B have coefficients of

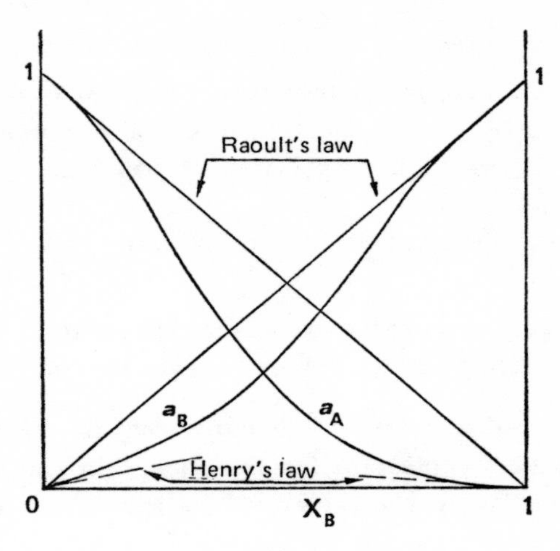
1
1
Raoult's law
a_B
a_A
Henry's law
0
X_B
1

FIG. 1

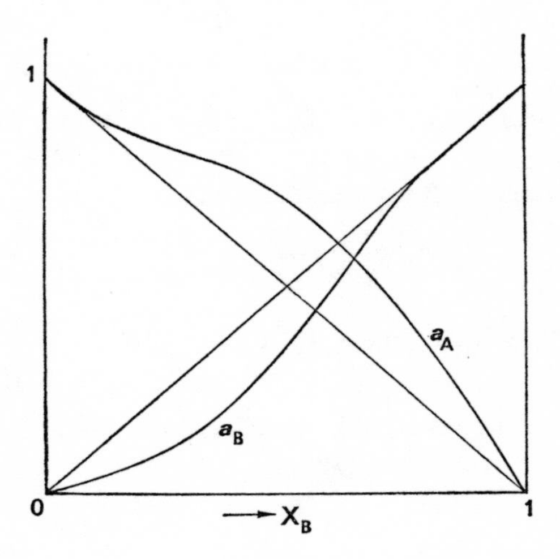
1
a_A
a_B
0
$\longrightarrow X_B$
1

FIG. 2

activity less than unity. In Figure 2 element B, in dilute solution in A, has a coefficient γ smaller than unity, and element A, in dilute solution in B, has a coefficient γ greater than unity.

A and B obey Raoult's law as solvents.

3.5. Concentrated Non-Ideal Solutions. Analytical Expression for Free Energy

For a non-ideal solution it is convenient to construct a mathematical expression for the change in free energy which enables us to make use of the results obtained by experiment. We use the coefficients of activity in the expression for the free energy:

$$\Delta G = RT \sum_i X_i \log \gamma_i + RT \sum_i X_i \log X_i .$$

The sum $RT \sum_i X_i \log \gamma_i$ may be condensed and represented by a term ΔH_m, which is expressed as a function of the X_i's.

For example, for a binary solution:

$$\Delta H_m = X_A X_B \left[\sum_0^j A_n (X_A - X_B)^n \right]$$

$$\Delta H_m = X_A X_B [A_0 + A_1 (X_A - X_B) + A_2 (X_A - X_B)^2 + \cdots] .$$

The coefficients A_n are, by definition, independent of X, but may depend on T and on P.

Amongst the non-ideal solutions we find those known as simple, where the coefficient A_0 is the only non-zero one:

$$\Delta H_m = X_A X_B A_0 .$$

These latter types of solution are important since after ideal solutions their behaviour is the simplest, both from the mathematical and the physical point of view, and the properties of some real solutions may correspond in practice to this formula.

Regular solutions

When A_0 is independent of temperature, the solution is regular. We will give further consideration to this type of solution, which plays a large part in statistical thermodynamics.

We will then put $A_0 = \mathcal{N}\omega$.

Subregular solutions

A number of empirical equations for non-ideal solutions where A_0 is not constant, have been suggested. Consideration has been given, for example, to a regular solution where the interaction term depended linearly on concentration and on temperature:

$$A_0 = (A_1 X_A + A_2 X_B)$$

$$\Delta G = A_1 X_A^2 X_B + A_2 X_A X_B^2 + RT(X_A \log X_A + X_B \log X_B) .$$

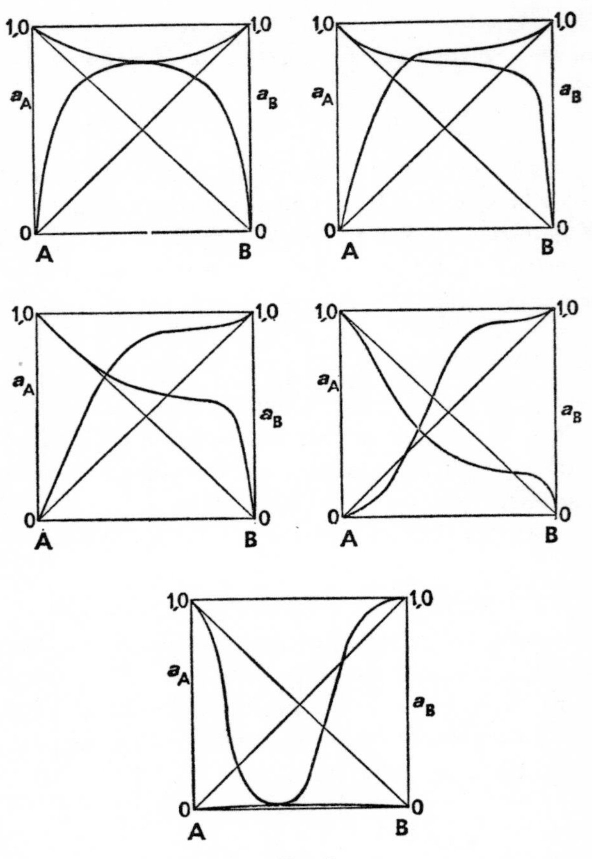

Fig. 3

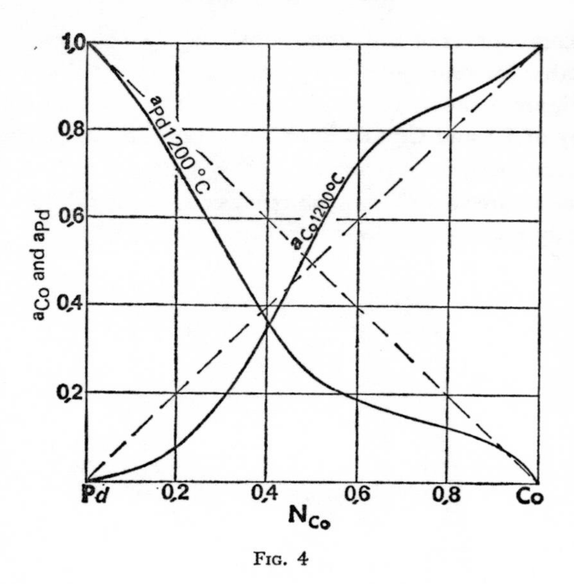

FIG. 4

This leads to the following chemical potentials:

$$\bar{G}_1 = \mu_1 = \mu_1^0 + RT \log X_A + \\ + X_B^2(2A_1 - A_2) + X_B^3(2A_2 - 2A_1)$$

$$\bar{G}_2 = \mu_2 = \mu_2^0 + RT \log X_B + \\ + X_A^2(2A_2 - A_1) + X_A^3(2A_1 - 2A2),$$

and to a semi-empirical expression for the coefficients of activity.

According to the values of A_1 and A_2, we can obtain

activity as shown in Figure 3, which are
entation of experimental results.
the experimental activity curves for an
Co at 1200° C. The corresponding curves
mixture are of very special form and they
y from the curves for ideal solutions or
olutions.

CHAPTER 4

EXPERIMENTAL DETERMINATION OF CURVES OF ACTIVITY AND OF FREE ENERGY

4.1. Measurement of the Activity of the Constituents Starting from Electromotive Forces (Electrolysis)

It is possible to construct a reversible electric cell, the electrodes of which are made from pure metal on one hand, and the solution on the other hand; if the solvent is less electropositive than the solute, we select as electrolyte a mixture of molten salts containing positive ions of the solute. The electrical energy liberated is equal and of opposite sign to the change in free energy corresponding to the following reaction:

$$\text{pure metal } A + \text{alloy} \rightarrow \text{solution in alloy};$$

for one mole of element A:

$$\Delta\bar{G}_A = \bar{G}_A - \bar{G}_A^0 = -\, nF\varepsilon,$$

where $\bar{G}_A$ is the partial molar free energy of constituent A, n is the valency of metal A, F is Faraday's constant, and ε is the electromotive force of the cell so constituted.

If a_A is the activity of metal A in the solution, converted to the pure state, we have:

$$-\, nF\varepsilon = RT \log a_A .$$

For the cell to operate under reversible conditions, we use a zero method which avoids any Joule effect.

The experimental conditions, temperature and nature of the electrolyte, are dependent on the elements of which the activities are being measured. This method can only be used in so far as a suitable electrolyte is available.

If we take measurements at different temperature, we can deduce from then the values of $\Delta\bar{S}_A$ and of $\Delta\bar{H}_A$.

We have, in fact, from relation (3):

$$\Delta\bar{S}_A = \bar{S}_A - \bar{S}_A^0 = nF\,(\partial\varepsilon/\partial T)_P\,,$$

and from relation (4):

$$\Delta\bar{H}_A = \bar{H}_A - \bar{H}_A^0 = nF\,[T\,(\partial\varepsilon/\partial T) - \varepsilon]\,.$$

4.2. Measurement of Vapour Pressure

Since, as we have seen, the activity is directly connected with the vapour pressure of the element in solution, measurement of this pressure leads directly to the activity in so far as the vapour may be regarded as a perfect gas.

In order to measure the partial pressure of the vapour it is necessary to arrange for favourable conditions. For example, one of the constituents must be only slightly volatile. We then use a manometer. Such a method has been used for solutions in mercury.

We can still measure the boiling point of the solution or the dew point of the vapour of the pure metal in equilibrium with the solution or with the alloy. A variant of this method makes use of thermal gravimetry to determine the precise moment at which equilibrium has been reached.

Finally, one method, frequently used but very delicate, is that of transfer by an inert gas or effusion.

4.3. Calorimetry

This is a method which has a tendency to increase in popularity. It consists of measuring the heat of solution of the alloy in a liquid metal or in an acid, and of comparing this value with the heat of solution of the constituent elements.

We then have the reaction:

Solid solution $(X_A X_B) \rightarrow A$ and B in the liquid metal or in the acid

$$\Delta H_1 = C_P \Delta T_1 ,$$

where H_i is the heat of reaction and C_p is the specific heat of the system.

With mechanical mixing of the constituents:

$X_A A + X_B B \rightarrow A$ and B in the liquid metal or in the acid

$$\Delta H_2 = C_P \Delta T_2 .$$

For the heat of formation of the solution:

$X_A A + X_B B \rightarrow$ solid solution (X_A, X_B)

$$\Delta H_m = (\Delta T_2 - \Delta T_1)\, C_p .$$

Unfortunately, this method is applicable only to a limited number of systems.

Equilibrium diagrams may also be used.

4.4. Determination of the Activity of Element B Starting from the Activity of Element A

If the activity a_A of element A is alone determined by experiment, the activity of element B is found with the aid of the Gibbs–Duhem relationship:

$$X_A \, d\bar{G}_A + X_B \, d\bar{G} = 0$$

$$X_A \, d \log a_A + X_B \, d \log a_B = 0,$$

and by graphical integration:

$$\log a_B = - \int_{a_A{}^0}^{a_A} \frac{X_A}{1 - X_A} \, d \log a_A .$$

This method of determination is difficult, however, when a tends to zero, since $\log a_A$ tends to infinity. Graphical integration is then no longer possible. We can, however, reach a result by consideration of the activity coefficients in place of the activities. We have in fact:

$$a_i = \gamma_i X_i$$

$$X_A \, d \log \gamma_A + X_B \, d \log \gamma_B + X_A \, d \log X_A + \\ + X_B \, d \log X_B = 0$$

$$dX_A = - \, dX_B \quad \text{and} \quad X_A \, d \log \gamma_A + X_B \, d \log \gamma_B = 0 .$$

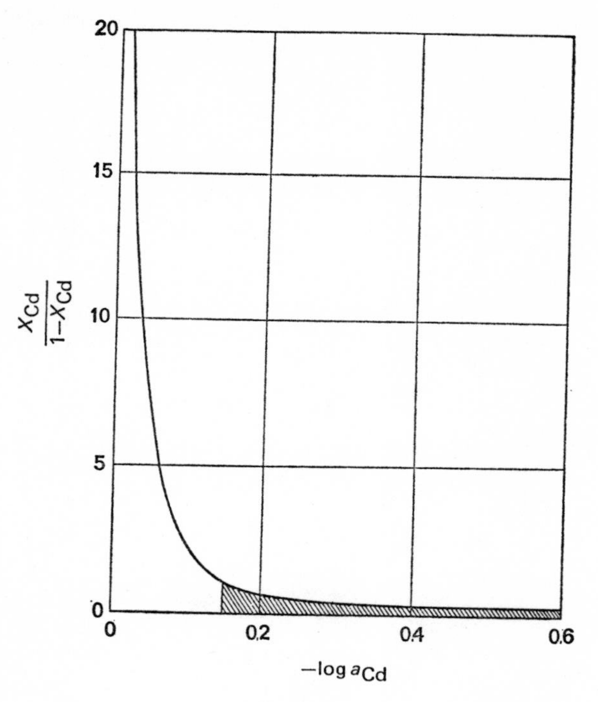

FIG. 5 *a*

γ_A is deduced from a_A and it suffices to integrate:

$$\log \gamma_B = -\int_{\gamma_A^0}^{\gamma_A} \frac{X_A}{1 - X_A} \, \mathrm{d} \log \gamma_A .$$

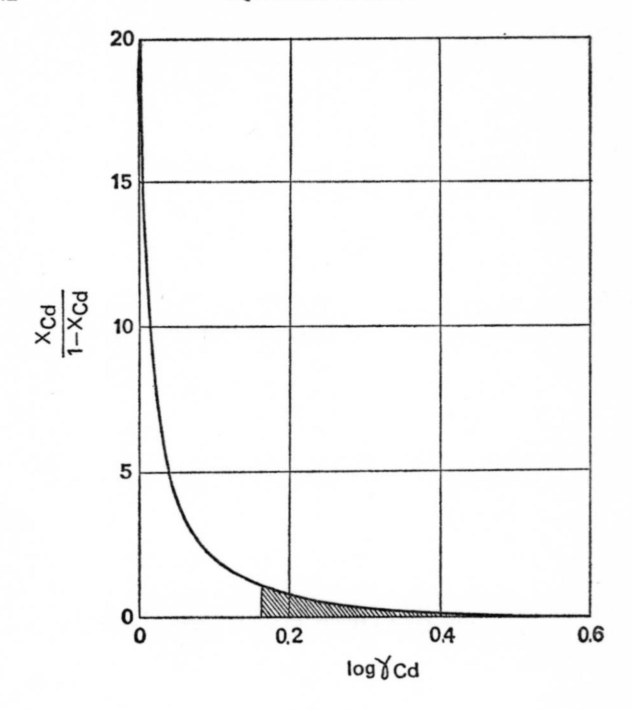

FIG. 5 *b*

For the upper limit of integration it is necessary to take into consideration that $\gamma_B \to \gamma_B^0$ in accordance with Henry's law, and that $\gamma_A \to 1$.

Figures 5a, b illustrate this reasoning for the example of solution of cadmium in lead.

CHAPTER 5

CONDITIONS FOR DE-MIXING OF SOLUTIONS

5.1. General Considerations

The expression for Gibbs' free energy per gram-atom for the ideal solution:

$$G = \bar{G}_A^0 X_A + \bar{G}_B^0 X_B + RT(X_A \log X_A + X_B \log X_B) = G_0 + \Delta G_m,$$

where:

$$G_0 = \bar{G}_A^0 X_A + \bar{G}_B^0 X_B,$$

shows that at constant pressure and composition the solid solution will always be more stable than any other combination such as, for example, the mixture of the pure constituents by mechanical means. In fact, the curve ΔG_m of excess free energy of mixture takes the form shown in Figure 6 whatever the temperature, always concave upwards. This is because the entropy of mixing is always positive, whence the second term of the preceeding expression is always negative. Thus there is no possibility of de-mixing whatever the temperature:

$$\Delta G_m = RT[X_A \log X_A + X_B \log X_B] \leqslant 0.$$

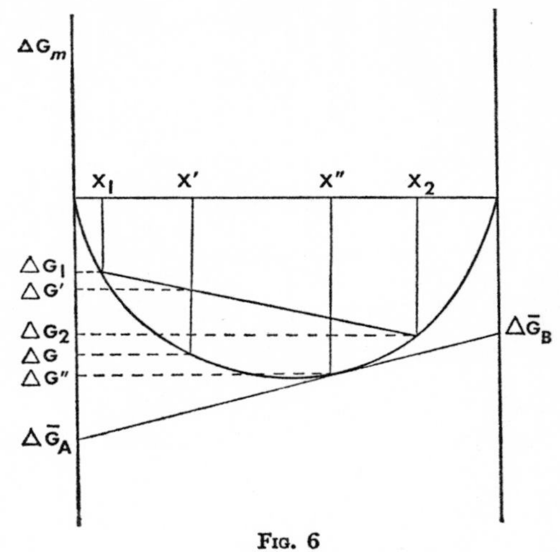

FIG. 6

This property is easily verified on Figure 6. Suppose that we have a composition X', the free energy of the ideal solution is given by the point ΔG. If we consider a mixture of two phases of composition X_1 and X_2, of which N_A and N_B are the constituents, if N_1 and N_2 are the numbers of atoms in solutions 1 and 2, we shall have:

$$X_1N_1 + X_2N_2 = XN = X(N_A + N_B)$$
$$(1 - X_1)\,N_1 + (1 - X_2)\,N_2 = (1 - X)\,N\,.$$

Further, for mechanical mixture of the solutions X_1 and X_2 we have:

$$N\Delta G' = N_1\Delta G_1 + N_2\Delta G_2\,,$$

and expressing N_1 and N_2 as functions of X, X_1 and X_2:

$$\Delta G' = \frac{X - X_2}{X_1 - X_2} \Delta G_1 + \frac{X_1 - X}{X_1 - X_2} \Delta G_2$$

$$(X_1 - X_2)(\Delta G' - \Delta G_1) = (X - X_2)\,\Delta G_1 - \\ - (X_1 - X_2)\,\Delta G_1 + (X_1 - X)\,\Delta G_2 .$$

After all calculations have been made, let:

$$\frac{\Delta G' - \Delta G_1}{\Delta G_2 - \Delta G_1} = \frac{X_1 - X}{X_1 - X_2} .$$

This shows that $\Delta G'$ lies on the chord joining the points ΔG_1, X_1 and ΔG_2, X_2 and that in fact the homogeneous solution corresponds to the most stable state, since ΔG is smaller than $\Delta G'$ whatever may be the values of X_1 and X_2.

5.2. Regular Solutions

Let us now consider the case of a regular solution where the excess of Gibbs' free energy per gram-atom of solution is given by:

$$\Delta G_m = \mathcal{N} X (1 - X)\,\omega + \\ + RT\,[X \log X + (1 - X) \log (1 - X)] ,$$

where $\omega > 0$.

This expression contains two terms, the first of which is the enthalpy of mixture and the second the entropy. We note that the entropy term is equal to that for ideal solutions.

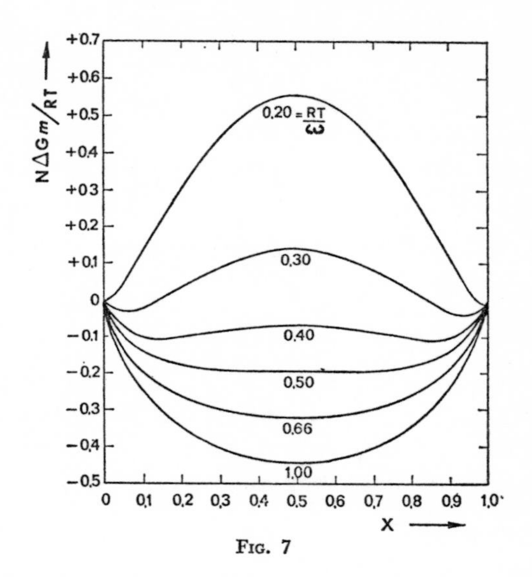

FIG. 7

If ω depends neither on temperature nor on concentration, the form of ΔG_m varies as a function of T.

The enthalpy and entropy terms have effects which oppose each other and may be more shortly written:

$$\Delta G_m = \Delta H_m - T\Delta S_m .$$

If we draw the curves for decreasing temperatures, we find a progressive change in shape which is shown on Figure 7, on which the values of

$$N\Delta G_m/RT$$

are shown for different values of RT/ω.

The highest curves correspond to the lowest temperatures. They show two minima which disappear when $RT/\omega=0.5$; the temperature T_c above which the curve is always concave upwards, is called the critical temperature.

The reason for the appearance of the minima at low temperatures is a consequence of the opposing effects of the terms ΔH_m and ΔS_m.

For very dilute solutions the entropy term always preponderates; it imposes the direction of change on curve $\Delta G_m(X)$, but for high concentrations the positive enthalpy term is greater and reverses the direction of change.

In consequence of the symmetry of behaviour of the two elements A and$B$, we find that the curves are symmetrical with respect to $X=0.5$.

This particular form of the free energy curve is of great interest since it shows that the homogeneous solid solution of elements A and B is not always the most stable combination at low temperatures. In this connection consider Figure 8 and take a composition X' between 0 and X_α corresponding to the first minimum. For this composition X' the homogeneous solution is more stable than any mixture of solutions obtained by drawing a chord through two points on the rising branches on one side and the other of X'. This brings us back to the case of the ideal solution already considered. The dilute solid solution behaves very much as an ideal solution.

On the other hand, take a composition X between X_α and X_β; the ΔG_m corresponding to the homogeneous solu-

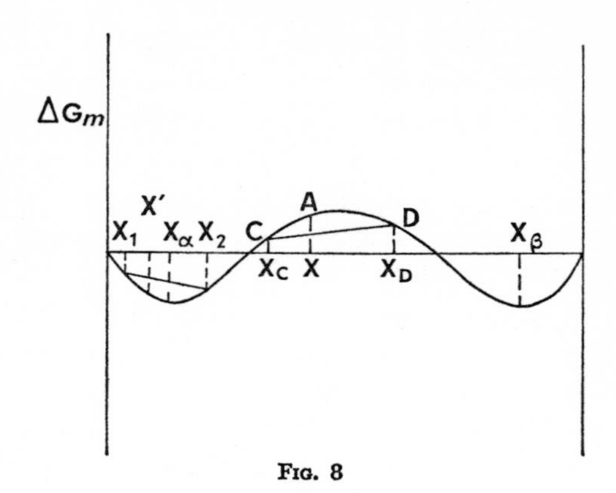

Fig. 8

tion is on the curve and is represented by point A. For the same X, however, any mixture of two phases with compositions given by the intersections C and D of a chord with curve corresponds to a smaller free energy. The total free energy is in fact $\Delta G'$ corresponding to the mixture of phases of composition X_C and X_D vertically below points C and D.

It is easily seen that the free energy is at a minimum for the mixture of solutions α and β; their compositions X_α and X_β are on the common tangent to the two minima. The algebraic calculation given later, starting from the expression for ΔG_m, shows that the graphical reasoning is exact.

In the case of a non-regular solution, the curves ΔG_m

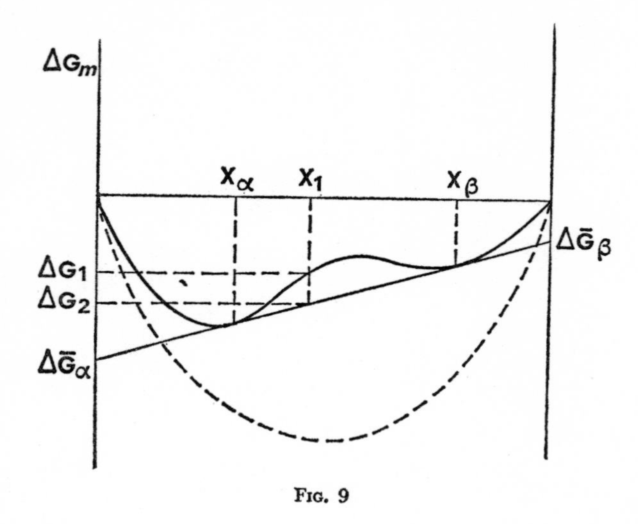

FIG. 9

are no longer symmetrical with respect to $X=0.5$, but assume forms comparable to those of Figure 9.

In this case, for concentrations between X_α and X_β (defined by the points of contact of the common tangent to the two arcs convex upwards), the mixture of the two phases of concentration X_α, $(1-X_\alpha)$ and X_β, $(1-X_\beta)$ is more stable than the homogeneous solution, below a certain temperature. Above this temperature there is only one single minimum and the homogeneous solution is the most stable combination for all concentrations (curve shown by broken line).

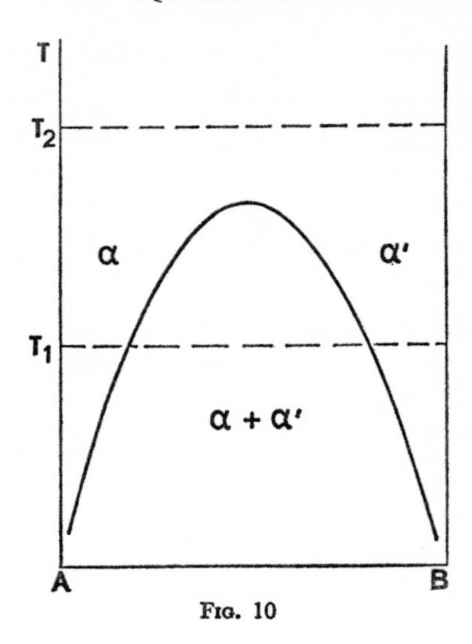

FIG. 10

We would stress that the position of the points X_α and X_β varies with temperature, as in the case of the regular solution.

Hence a knowledge of the free energy curves at all temperatures enables us to predict the existence and type of phases which will be present at a given temperature and for a given composition.

Since the curves $\Delta G_m(X)$ vary with temperature it is, however, simpler to use another type of diagram.

For a system A–B we can have a homogeneous solution at high temperature and a mixture of two phases at low temperature. We then draw what is know as an equilibrium diagram which shows the positions of points X_α and X_β as functions of the temperature. In practice we plot the temperature T as the ordinate and the concentration as the abscissa. Figure 10 shows such a diagram.

At low temperatures (T_1) we find two terminal solid solutions separated by a two-phase domain ($\alpha + \alpha'$); at high temperatures there is one domain only, comprising a continuous homogeneous solid solution of α in α'.

5.3. Determination of Lines of Transformation

This being so, we must determine the relationships which express the equilibrium of the two phases α and α'.

We must seek the minimum value of ΔG_m corresponding to the set of equations:

$$\Delta G_m = \Delta G_\alpha + \Delta G_\beta$$

and

$$\Delta G_m = \sum \Delta \bar{G}_{i_\alpha} X_{\alpha_i} + \sum \Delta \bar{G}_{i_\beta} X_{\beta_i} .$$

We have shown above that the activity of a constituent is determined by the saturated vapour pressure above the solution. If two phases, liquid or solid, are in equilibrium, the vapour above them is composed of the different constituents the partial pressures of which are well defined and unique. This therefore leads to one unique activity for each constituents in all phases.

The partial molar free energy of each constituent is the same in two phases in equilibrium.

If we consider Figure 6 again, we see that for any X:

$$\Delta G_m = (1 - X)\,\Delta \bar{G}_A + X \Delta \bar{G}_B = \Delta G_m(X)\,.$$

This is one for mole of alloy A–B.

For n moles we would have:

$$n\Delta G_m = n_A \Delta \bar{G}_A + n_B \Delta \bar{G}_B = n\Delta G_m(X)\,.$$

Putting $\Delta G_m = f$:

$$\Delta \bar{G}_B = \frac{\partial (n\Delta G_m)}{\partial n_B} = \frac{\partial n}{\partial n_B} f + n \frac{\partial f}{\partial n_B} =$$

$$= f + n \frac{\partial f}{\partial X} \frac{\partial X}{\partial n_B} = f + n \frac{\partial f}{\partial X} \frac{n_A}{n^2}$$

$$\Delta \bar{G}_B = f + \frac{\partial f}{\partial X}(1 - X)$$

$$\Delta \bar{G}_A = f - \frac{\partial f}{\partial X} X\,.$$

It is easily seen that these values are obtained directly by taking the tangent at the point $(\Delta G_m, X)$ and its intersections with the ordinate axis corresponding to $X = 1$ and $X = 0$.

If we have two phases α and β in equilibrium corresponding to compositions X_α and X_β (Figure 9), we find these compositions by putting the values of $\Delta\bar{G}_B$ and of $\Delta\bar{G}_A$ respectively equal in the two phases:

$$f_\alpha + \frac{\partial f_\alpha}{\partial X_\alpha}(1 - X_\alpha) = f_\beta + \frac{\partial f_\beta}{\partial X_\beta}(1 - X_\beta)$$

$$f_\alpha - \frac{\partial f_\alpha}{\partial X_\alpha} X_\alpha = f_\beta - \frac{\partial f_\beta}{\partial X_\beta} X_\beta .$$

By comparing terms:

$$\frac{\partial f_\alpha}{\partial X_\alpha} = \frac{\partial f_\beta}{\partial X_\beta} .$$

This shows analytically that the tangents to the curve $G(X)$ at X_α and X_β must coincide. In order to find the concentrations of two phases in equilibrium it is sufficient to take the common tangent to the two parts of the curve $\Delta G_m(X)$.

5.4. Case of the Strictly Regular Solution

In the case of the strictly regular solution we have per mole of alloy:

$$\Delta G_m = X(1 - X) A_0 + + RT [X \log X + (1 - X) \log(1 - X)] ,$$

where $A_0 > 0$. Differentiating with respect to X:

$$\frac{\mathrm{d}(\Delta G_m)}{\mathrm{d}X} = (1 - 2X) A_0 + RT \left[\log\left(\frac{X}{1 - X}\right)\right] .$$

For ΔG_m to be a minimum, $\mathrm{d}\Delta G_m/\mathrm{d}X$ must $=0$:

$$(1-2X)\,A_0 + RT\left[\log\frac{X}{1-X}\right] = 0\,.$$

We can solve this equation by seeking the intersection of the straight line

$$y = A_0(2X-1)$$

and the curve:

$$y = RT\log\frac{X}{1-X} \qquad 0 \leqslant X \leqslant 1\,.$$

First note that the equation is identically zero for $X=0.5$; hence $y=0$. $X=0.5$ is one of the points sought. It corresponds to the maximum of the curve ΔG_m.

If we draw the straight line and the curve, we see that they intersect in two points which are symmetrical with respect to $X=0.5$. These are true minima corresponding to the phases α and α'.

If the temperature rises, the two points of intersection move towards the point $y=0$, $X=0.5$. They coincide when we enter the single-phase domain. We then have:

$$T_c = A_0/2R = \omega/2k\,.$$

The existence of a separation shown by the presence of two minima on the free energy curve is also demonstrated by the form of the activity curves.

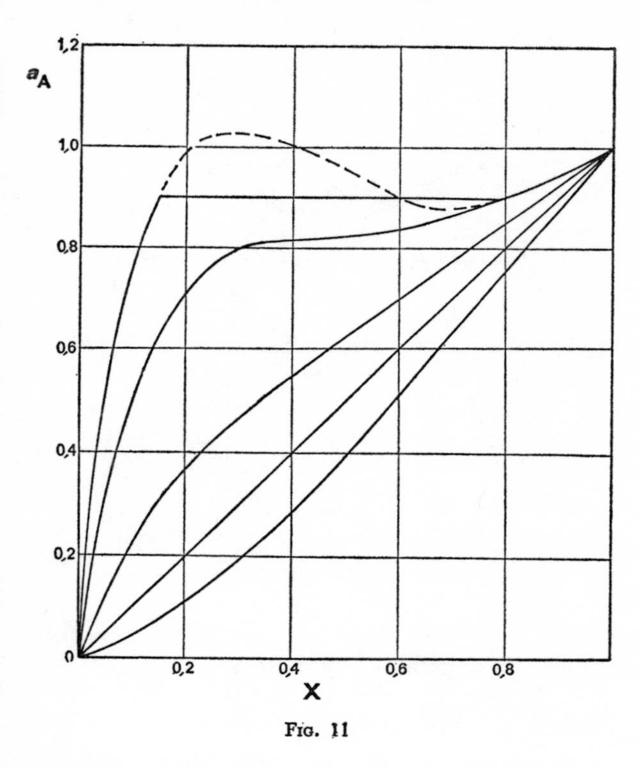

FIG. 11

Figure 11 show some *theoretical examples of activity curves* for different temperatures. When the curves have turning points, as in the case of the upper curve, there is separation into two phases. The activity corresponding to

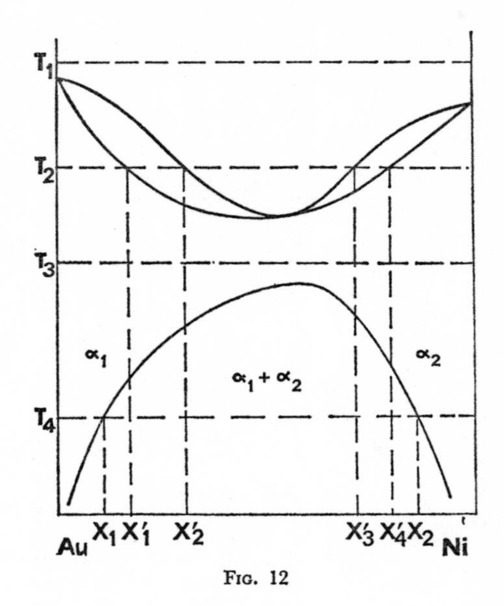

FIG. 12

equilibrium is determined by putting:

$$a_{\alpha_A} = a_{\beta_A} \quad \text{and} \quad a_{\alpha_B} = a_{\beta_B}.$$

As an example, consider the equilibrium diagram of the *Au–Ni system* (Figure 12). It comprises three domains:

- at high temperatures, a continuous liquid homogeneous solution;
- below this, a domain where a liquid phase exists in the presence of a solid phase;

- at medium temperatures, a continuous homogeneous solid solution;
- at low temperatures, two terminal solid solutions α_1 and α_2.

Thus, in the solid state and for a given temperature, we may have a homogeneous phase the composition of which varies continuously with the concentration of the constituents of the alloy, or a mixture of phases of which the composition does not vary but of which the relative volume varies with the overall concentration of the alloy. This indicates the variance of the system, which is controlled by the phase rule which we will now state. This rule, very useful whenever more than two constituents enter into the system, is only valid at constant pressure.

5.5. Phase Rule

Let c be the number of constituent elements distributed over p phases. The composition of each one of the phases is defined by $c—1$ concentration terms. The number of variables due to concentration is $p(c—1)$. Since pressure and temperature can also vary, there are in all $2+p(c—1)$ variables.

We have seen above that the chemical potential of a constituent is the same in all phases in equilibrium.

$$\mu_A^\alpha = \mu_A^\beta = \mu_A^\gamma = \ldots$$

$$\mu_B^\alpha = \mu_B^\beta = \mu_B = \ldots$$

$$\ldots\ldots\ldots\ldots\ldots\ldots$$

There are therefore $(p-1)$ independent equations for each constituent, hence in all $c(p-1)$ relationships.

These relationships reduce the number of independent variables to such an extent that the variance v of the system is equal to the total number of variables less the number of relationships connecting them:

$$v = p(c-1) + 2 - c(p-1)$$

$$\boxed{v = c - p + 2\,.}$$

In the case of a binary alloy, *at constant temperature and pressure*, there are only two constituents:

$$v = 2 - p\,.$$

If $p=1$, $v=1$, there is one variable only, the concentration.

If $p=2$, $v=0$, the concentration is fixed in each phase.

In general, in metallurgy, we vary the temperature only. The variance is reduced by one. However, it is sometimes necessary to have regard to the pressure, and the variance then becomes normal.

The phase rule is very useful for the study of equilibrium diagrams, since it gives the number of phases which can exist together.

CHAPTER 6

EQUILIBRIUM DIAGRAMS

We will commence by the diagram for binary alloys forming quasi-ideal continuous solutions in the solid and liquid states; we will then describe the diagrams for more complicated cases where the properties of the elements are very different.

6.1. Alloys Forming Ideal Solutions in the Liquid State and in the Solid State but which Have Differing Melting Points

Let two elements be soluble in all proportions in the liquid or solid state. The equilibrium diagram contains curves only, the loci of points where the liquid and the solid are in equilibrium. To determine these curves we must calculate the free energy corresponding to each of the phases. The general expression for free energy of solutions leads, for the solid and the liquid, to the two following differential relationships:

$$\mathrm{d}G^S = V^S\,\mathrm{d}P - S^S\,\mathrm{d}T + \bar{G}_A^S\,\mathrm{d}X_A^S + \bar{G}_B^S\,\mathrm{d}X_B^S$$
$$\mathrm{d}G^L = V^L\,\mathrm{d}P - S^L\mathrm{d}T + \bar{G}_A^L\mathrm{d}X_A^L + \bar{G}_B^L\mathrm{d}X_B^L.$$

If we consider *constant pressure* conditions, and if the

liquid is in equilibrium with the solid, we have:

$$dG^S = dG^L$$
$$\bar{G}_A^L = \bar{G}_A^S = \bar{G}_A \quad \text{and} \quad \bar{G}_B^L = \bar{G}_B^S = \bar{G}_B.$$

from which:

$$-(S^S - S^L)\,dT = \bar{G}_A(dX_A^L - dX_A^S) + \\ + \bar{G}_B(dX_B^L - dX_B^S).$$

Now:

$$X_A + X_B = 1 \quad dX_B^L = -dX_A^L \quad \text{and} \quad dX_B^S = -dX_A^S,$$

from which:

$$(S^L - S^S)\,dT = (\bar{G}_A - \bar{G}_B)(dX_A^L - dX_A^S).$$

Since we are dealing with an ideal solution, we might think that the liquid is in contact at all temperatures with solid of the same composition, i.e.

$$X_A^L = X_A^S,$$

whatever the temperature. This is not true because the melting points are different.

By making a slight change in the composition of the alloy, dX, we would have:

$$dX_A^L/dX = dX_A^S/dX \quad \text{whence} \quad dT/dX = 0$$

whatever the value of X.

But since $T_{f_A} \neq T_{f_B}$, dT/dX is not zero; therefore the line separating solid from liquid is not straight.

The phase rule, in fact, requires that the liquid and solid phases should be separated by a two-phase domain. The passage from the liquid state to the solid state cannot be sudden.

Let us now establish the limits of this two-phase domain. We can calculate them from the expressions for the free energy fo the two phases present.

We choose the pure solids as a reference state. Under these conditions, we put for the ideal liquid and solid solutions:

$$\Delta G_m^L = RT\,[X_A^L \log X_A^L + X_B^L \log X_B^L + X_A^L \Delta G_{fA} + X_B^L \Delta G_{fB}] \text{ for liquid state.}$$

ΔG_{fA} and ΔG_{fB} are the variations in molar free energy of the constituents in the pure state when they pass from solid to liquid, and

$$\Delta G_m^s = RT\,[X_A^S \log X_A^S + X_B^S \log X_B^S]$$

for the solid state.

The values of ΔG_{fA} and ΔG_{fB} depend on the temperature. In particular, these terms are negative when the temperature is above the melting point and positive below this.

Consider one of the pure constituents at the melting point T_{f_i}.

We have $\Delta G_{f_i} = \Delta H_{f_i} - T_{f_i} \Delta S_{f_i}$. Since at the melting point the liquid is in equilibrium with the solid:

$$\Delta G_{f_i} = 0, \quad \text{whence} \quad S_{f_i} = \Delta H_{f_i} / T_{f_i}.$$

At temperature T different from T_{f_i} we would have:

$$\Delta G_{f_i} = \Delta H_{f_i} - T\Delta S_{f_i}.$$

In general, ΔH_{f_i} and ΔS_{f_i} are independent of the temperature.

We therefore deduce the relationship:

$$\Delta G_{f_i} = \Delta H_{f_i}(1 - T/T_{f_i}).$$

When melting takes place at a temperature different from T_{f_i}, the change in free energy is not zero.

This introduces into the expression for ΔG_m^L two terms which depend on temperature.

The different extreme cases are the following (Figure 13):

For $T = T_{f_B}$, the intersection of the curves is the point $X = 0$, $\Delta G_m = 0$ (Figure 13b).

For $T_{f_A} < T < T_{f_B}$, the curves intersect in one point which depends on the temperature (Figure 13c).

For $T = T_{f_A}$ the curves intersect in one single point T_B.

Lastly, for $T < T_A$ there is no common point and the solid becomes the most stable phase.

Let us consider what takes place between T_{f_B} and T_{f_A}.

We can draw a tangent common to the two curves, liquid and solid. Between the two compositions corresponding to the points where the tangent is common, X_B^L and X_B^S, we have a mixture of two phases: a liquid phase and a solid phase.

Let us seek the compositions X_B^L and X_B^S.

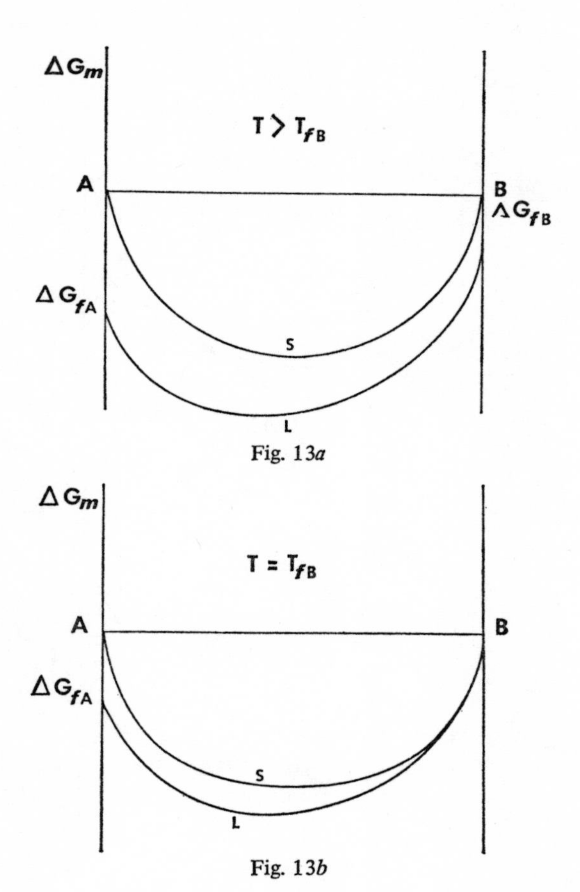

Fig. 13*a*

Fig. 13*b*

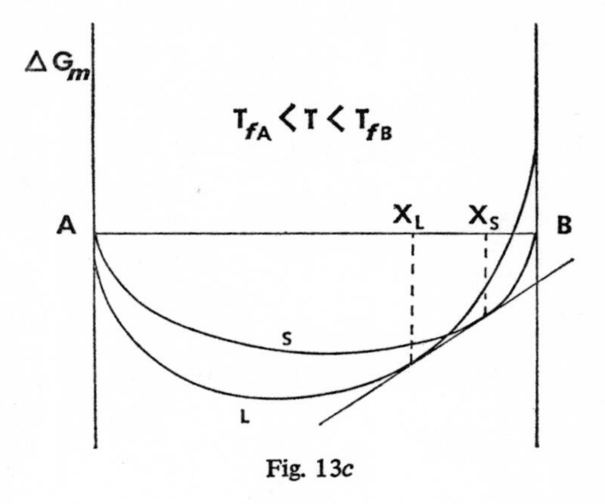

Fig. 13*c*

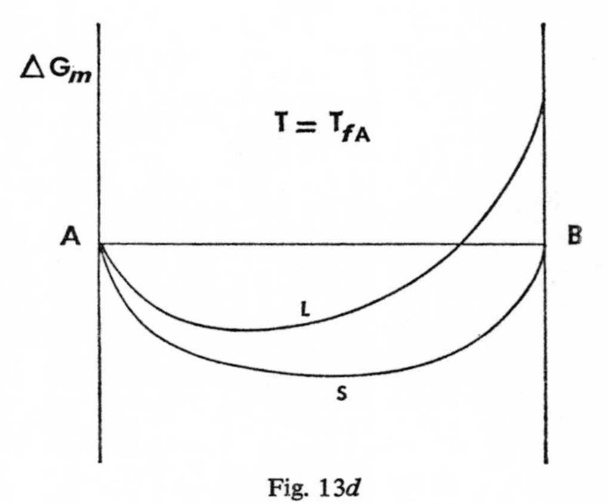

Fig. 13*d*

By putting the chemical potentials of A and B respectively the same in the liquid and in the solid in equilibrium, we find per mole:

$$\log(X_B^S/X_B^L) = \Delta G_{f_B}/RT \quad \text{and} \quad \log(X_A^S/X_A^L) = \Delta G_{f_A}/RT .$$

If we put

$$\exp(\Delta G_{f_A}/RT) = \alpha \quad \text{and} \quad \exp(\Delta G_{f_B}/RT) = \beta ,$$

we have

$$X_B^L = \frac{\alpha - 1}{\alpha - \beta}$$

$$X_B^S = \beta \frac{\alpha - 1}{\alpha - \beta}$$

$$X_A^L = \frac{1 - \beta}{\alpha - \beta}$$

$$X_A^S = \alpha \frac{1 - \beta}{\alpha - \beta} .$$

When we graph the values of X_B^L and X_B^S against temperature, we obtain the diagram shown in Figure 14. This diagram could correspond for example to the systems Ge–Si, Cu–Ni, Cu–Pd, Au–Ag, Ag–Pd and Ir–Pt.

The domain of coexistence of liquid and solid becomes more extensive as the difference between the melting points of the two alloys increases.

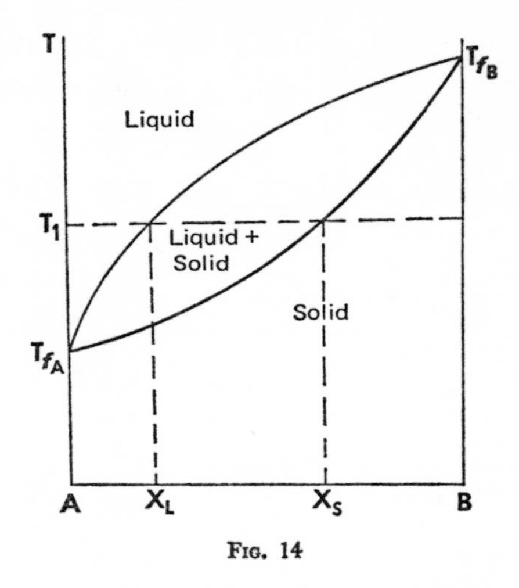

Fig. 14

The curves thus defined are called the solidus and the liquidus. We can easily verify that the phase rule holds good over the whole diagram.

This calculation can be applied to non-ideal solutions, provided that we know the expression for the free energy of mixture for the liquid and that for the solid. The equation of the solidus and of the liquidus is not so simple as for ideal solutions.

6.2. Alloys of Greatly Differing Elements

We will now see the general trend of the equilibrium diagrams that can be deduced from the free energy

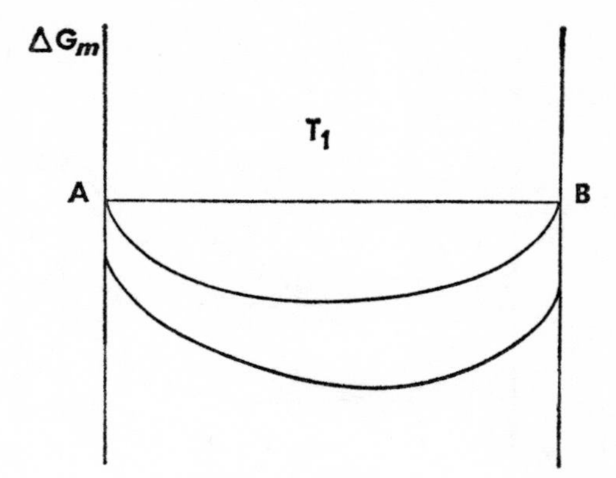

FIG. 15*a*

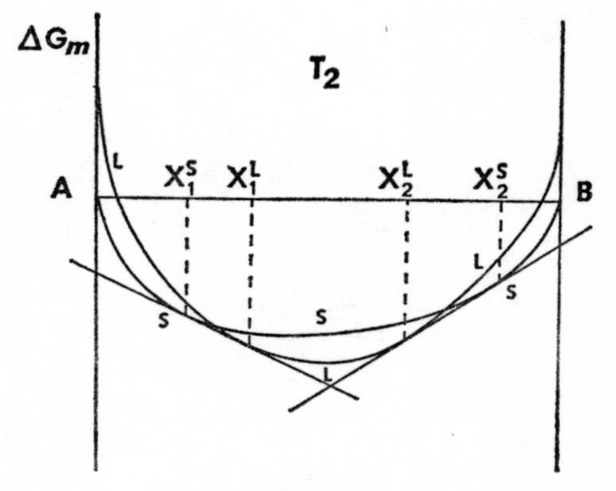

FIG. 15*b*

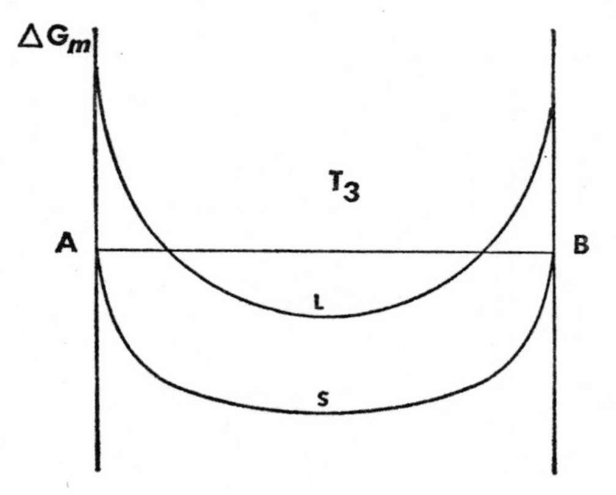

FIG. 15*c*

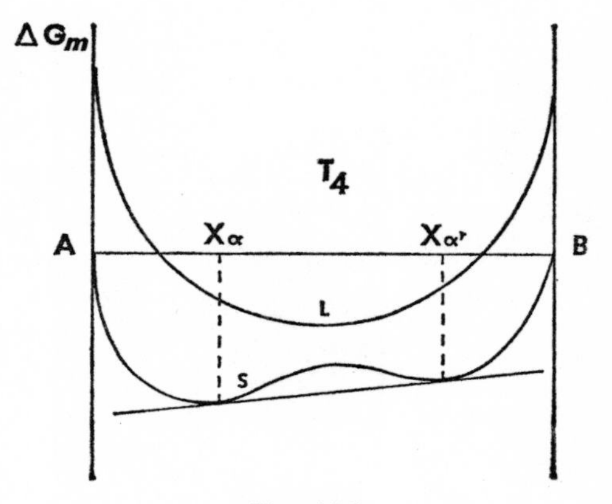

FIG. 15*d*

curves, without giving the details of the calculations.

Consider a binary system which, for three temperatures T_1, T_2, T_3 and T_4, exhibits the three types of free energy curve shown in Figures 15a, b, c, and d. Figure 15a corresponds to a high temperature T_1 and shows the theoretical free energy curves for the alloy in the solid state and in the liquid state. We see that the liquid phase is more stable than the solid phase, whatever the temperature.

On Figure 15b, on the other hand, the curves for the two states intersect in two points. At this temperature, T_2, we can have two solid solutions, two domains where there are two phases, solid and liquid, and one intermediate liquid solution.

At temperature T_3 we have the conditions shown in Figure 15c. The two free-energy curves do not intersect but at T_4 we find two solid solutions and a domain with two solid phases. This is also found on the equilibrium diagram of Figure 12, on which the temperatures T_1, T_2, T_3 and T_4 are marked.

This type of equilibrium diagram corresponds for example to the Au-Ni and Cu-Mn systems.

Figure 16 represents a type of diagram where the domain with two solid phases comes into contact with the liquid. Figure 17 represents the free energy curves corresponding to temperature T_2 of the diagram in Figure 16.

This type of diagram is known as a diagram showing a eutectic. When the two elements have different melting points, we also find peritectic diagrams such as the examp-

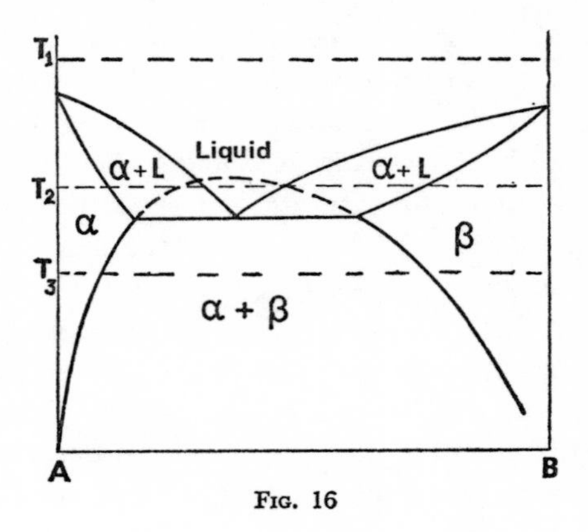

Fig. 16

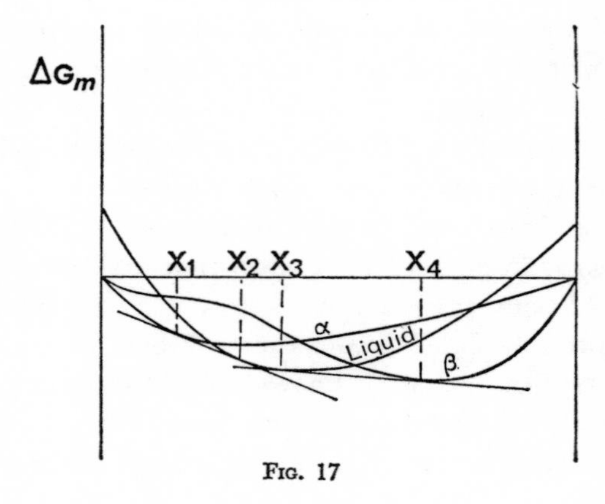

Fig. 17

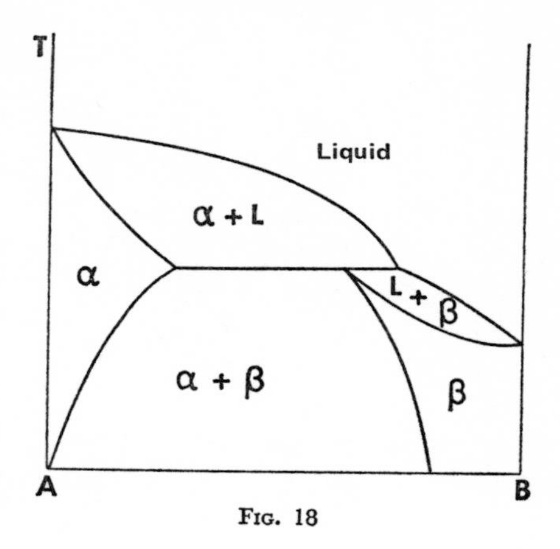

Fig. 18

le given in Figure 18. This corresponds to the alloys Cu-Co and Pt-W.

Finally, certain systems have more complicated equilibrium diagrams. In particular these diagrams may include domains of one single phase corresponding to the formation of intermetallic compounds as is shown by the example in Figure 19. The free energy curves applicable to temperature T_1 are shown on Figure 20.

For certain systems there may be more than two metallic compounds. These compounds may be stoichiometric, or may have a range of compositions extending in both directions from the theoretical composition.

For all types of the preceding diagrams we can verify the validity of the phase rule. When only one phase is

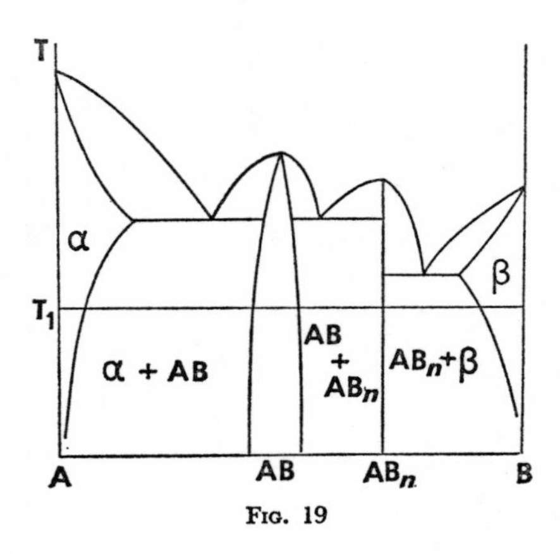

FIG. 19

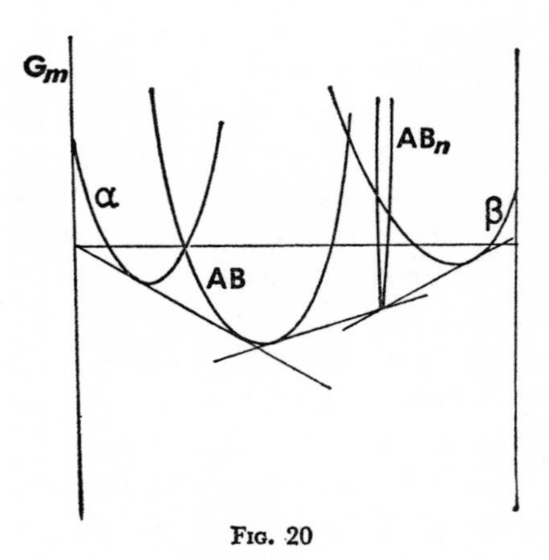

FIG. 20

present, there are three degrees of freedom; at constant temperature and pressure, the composition can change. If two phases exist together, there are two degrees of freedom: at constant temperature and pressure the composition is fixed for each of the phases. Finally, three phases can only exist together at one fixed composition and temperature, since the number of degrees of freedom is then zero. This is the case of the eutectic or the peritectic along

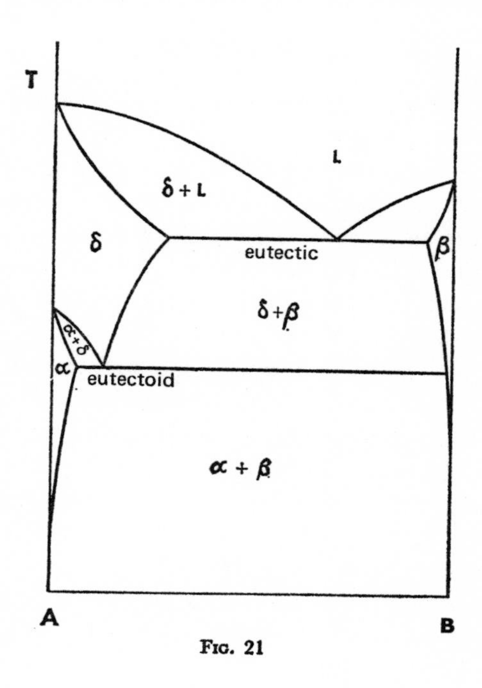

Fig. 21

which the liquid and the two solid phases have fixed compositions.

In the solid state we find reactions analogous to eutectic reactions; these are known as eutectoid reactions. Figure 21 shows the typical arrangement of the equilibrium diagram. We find this type of diagram in the iron-carbon system. The free energy curves have the same form as for the eutectic reaction, but the phases present are solid.

6.3. Ternary Alloys

The case of binary alloys that we have just discussed is relatively simple. When more than two elements enter into the composition of an alloy, there are more degrees of freedom and more than three phases can coexist in equilibrium.

It is still easy to represent equilibrium for such alloys. To do this we use sections of a three-dimensional diagram constructed on an equilateral triangle, the principle of which is shown on Figure 22.

Each apex of the triangle corresponds to one of the pure elements, and a point inside corresponds to a composition determined by dropping perpendiculars onto the sides from this point.

Thus for the composition of an alloy P we have:

concentration of A: $X_A = PP_1/AO_1$
concentration of B: $X_B = PP_2/BO_2$
concentration of C: $X_C = PP_3/CO_3$,

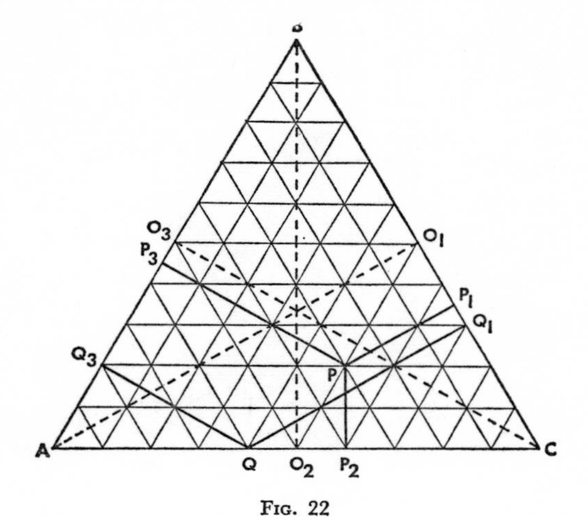

FIG. 22

and since $AO_1 = BO_2 = CO_3 = h$, we have:

$$X_A + X_B + X_C = 1\,.$$

If the concentration of one of the elements is zero, for example $X_B = 0$, we read the concentration of the two others directly from the corresponding side, at Q. The three sides are graduated, in this example at intervals of 0.1 or 10%.

The point representing the alloy is finally obtained by giving a height equal to the temperature under considera-

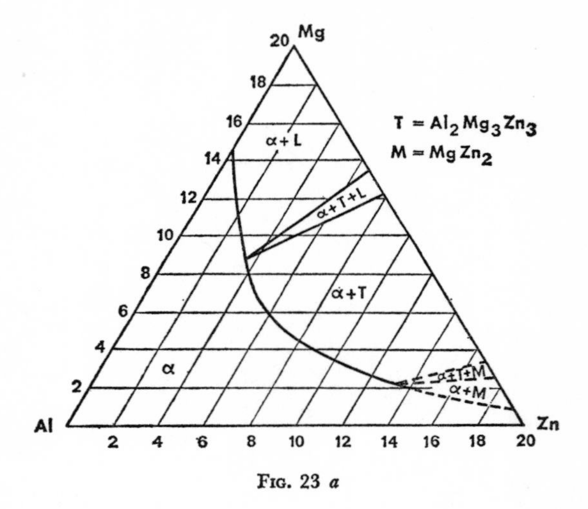

FIG. 23 *a*

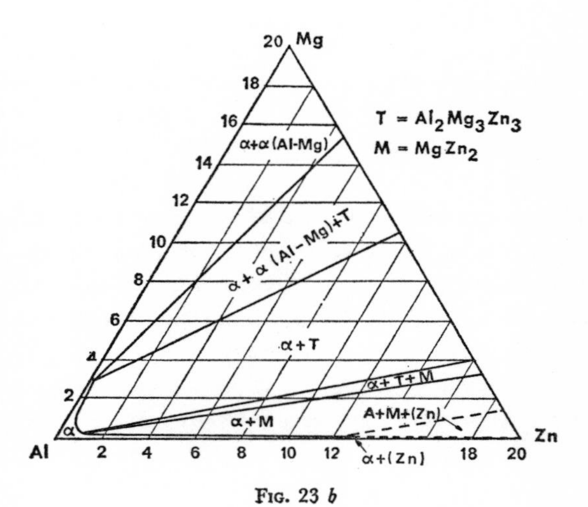

FIG. 23 *b*

tion. We thus trace transformation surfaces:

$$T = f(X_A, X_B, X_C).$$

Isothermal sections are often used on which are found the intersections with the transformation surfaces.

Figures 23a and b represent respectively the partial sections at 450°C and 100°C for the aluminium-zinc-magnesium system.

Knowledge of the equilibrium diagrams for more than two elements is not yet very extensive. It would, however, clearly be of use, since the majority of industrial alloys contain more than two metals.

6.4. Determination of Equilibrium Diagrams. Experimental Methods

From the preceding, one might imagine that equilibrium diagrams are determined from a knowledge of the free energy curves. Unfortunately, the facts are quite different. The free energy curves are very difficult to establish theoretically or experimentally and they are in fact most frequently deduced from the information on the equilibrium diagrams. To do this, it is sufficient to have available a large number of alloys the structures of which are determined as functions of the temperature. An equilibrium diagram is usually obtained progressively, in parts and by different researchers, and in most cases it represents the synthesis of a large number of investigations. These parts of the diagram are of interest only for particular practical problems. Certain laboratories however have adopted as their aim the systematic investigation of the diagrams of great utility.

In order to determine a diagram it is necessary to examine the equilibrium state for every composition, at constant pressure, at all temperatures. This equilibrium state is reached within a reasonable time only at relatively high temperatures. We have in principle a mixture of elements, say *A* and *B*, which we melt. During cooling, we determine the solidification point by thermal analysis. Then we maintain the solid alloy or the mixture of solid and liquid for a sufficiently long time for each phase to reach its equilibrium state. [1] We then quench the alloy and examine the sample obtained at room temperature.

Thermal analysis and analysis by expansion methods lead easily to information on the transformation lines. These methods are however misleading if the temperature is not high enough for equilibrium to be established. They are most frequently used as a preliminary attack on the problems.

In many cases microscopic examination is sufficient to determine the extent of the domains of one or two phases, since phases of different structure or composition do not exhibit the same micrographic appearance. To determine the positions of the domain limits, however, recourse must be had to physical methods, for greater precision.

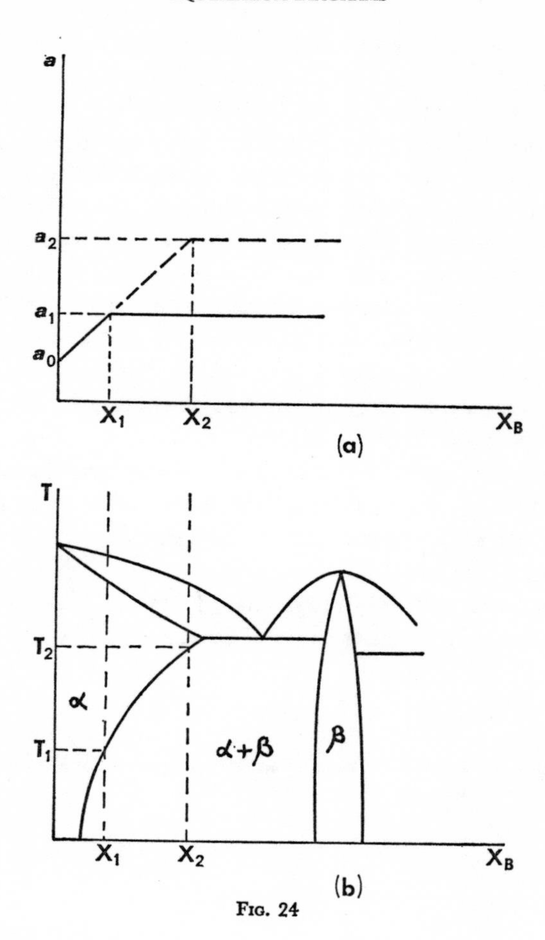

FIG. 24

For the limit of solubility in a terminal solid solution (Figure 24b), X-rays are the most precise method.

For a given temperature the parameters of the crystal lattices vary continuously with the concentration. When we reach the limit of solubility, we are in the two-phase region where the composition does not change, but the relative volume alters. The graph of the parameter against the composition therefore has the general form shown on Figure 24a. The concentration corresponding to the limit of solubility is then that where the tangent to the curve exhibits a discontinuity and no longer exists.

In many cases the precision is sufficient to give information about the position of the limit.

X-ray diffraction also gives information about the nature of the phases and enables intermediate phases to be identified; this is complemented for greater certainty by chemical analysis.

The great difficulty often arises from the impossibility in practice of reaching equilibrium for the alloy.

NOTE

[1] The equilibrium state is reached when maintenance for a longer time no longer produces a change in structure.

CHAPTER 7

STATISTICAL THERMODYNAMICS OF SOLUTIONS

7.1. General Considerations. Structural and Atomic Aspect

Everything up to now constitutes a classical approach to the thermodynamics of metallic solutions. It can appear formal and abstract without support from the structure of matter. But in fact, whilst atomic theory leads to identical conclusions, it enables the metallurgist to understand the mechanisms better and to link them with the crystallographic structure, of which he has a continually deepening knowledge.

As we have already said, metals do not exist in general in the form of single crystals, but of polycrystalline aggregates consisting of a large number of small crystals.

The atoms in these crystals, although displaced by oscillation movements due to thermal agitation, occupy positions or sites which are the intersections of a triply-periodic network. We will discuss this structural aspect further in the Part II. Let us be content for the present to consider the sites as the mean positions of the different types of atoms which form the alloy. The thermodynamic relationships previously given may be deduced from the distribution of the atoms on these sites.

7.2. Functions of Partition

Consider a crystal composed of a very large number of atoms of differing nature. According to quantum theory this crystal may have different energy levels. The time spent in each of these depends on their energy.

We will suppose that the temperature, the volume and the composition of the crystal are fixed. Boltzmann has shown that under these conditions the probability that such a system will be at an energy level E_r is proportional to the factor:

$$\exp(-E_r/kT),$$

where T is the absolute temperature and k is the universal constant known as Boltzmann's.

In order to determine a thermodynamic property in equilibrium of this system, it is necessary to find the mean of all the quantum states of the system, having regard to their statistical weight g_r, and also to the fact that several states are equivalent (we then say that they have degenerated).

The fraction of the time spent by the system in one state is then equal to:

$$f_r = \frac{g_r \exp(-E_r/kT)}{\sum_r g_r \exp(-E_r/kT)},$$

and for any thermodynamic property, ϕ, the mean $\bar{\phi}$ is equal to

$$\bar{\phi} = \frac{\sum_r g_r \phi_r \exp(-E_r/kT)}{\sum_r g_r \exp(-E_r/kT)}.$$

In other words, the mean is taken over all states E_r with ϕ_r dependent on r. The sum in the denominator is known as the 'partition function' and we will represent it conventionally by the letter Q:

$$Q = \sum_r g_r \exp(-E_r/kT).$$

From the preceding we can deduce the relationships between Q and the classical thermodynamic functions.

We can show that F defined as:

$$F = -kT \log Q$$

possesses all the thermodynamic properties of the free energy of Helmholtz.

Thus for pressure we have:

$$P = \frac{\sum_r P_r \exp(-E_r/kT)}{\sum_r \exp(-E_r/kT)},$$

and from Equation (6):

$$P_r = -\partial E_r/\partial v$$

$$P = \frac{\sum_r -(\partial E_r/\partial v) \exp(-E_r/kT)}{\sum_r \exp(-E_r/kT)} =$$

$$= kT(\partial \log Q/\partial v)_T = -(\partial F/\partial v)_T.$$

We thus return to one of the equations of classical thermodynamics.

In the same manner we can obtain the equation:

$$S = -(\partial E/\partial T)_v .$$

Knowing the detailed structure of the system or a model which represents it with sufficient accuracy, it is sufficient to determine the partition function in order to find the expression for the free energy F.

In our case, since we are dealing with solids, we can replace F by G, the Gibbs free energy at constant pressure, and put:

$$G = -kT \log Q ,$$

and deduce from this the partial thermodynamic functions that we have used above.

7.3. Calculation of The Partition Function

When we wish to calculate the partition function of a system, we must do so for all the degrees of freedom. We can agree that in many cases some are independent of the others. The energy is then expressed by a sum of terms and Q by a product of factors.

As far as alloys are concerned, we can consider the degrees of freedom relative to the position of the atoms and to their oscillations about this position. This then leads to the consideration of two types of energy: energy of configuration and acoustic energy. The energy of configuration corresponds to the energy which the system might possess if the atoms were frozen in their mean positions. The acoustic energy is that leading to the vibration of the atoms about their mean positions. In what follows we will suppose that the modes of vibration are independent of the distribution of the atoms over the sites. We then write:

$$Q = Q_{ac}\Omega,$$

where Q_{ac} is the partition function for the modes of vibration and Ω the partition function for configuration.

Under these conditions, the free energy of mixture per atom of two species may be written:

$$-\Delta G_m/kT = \log \Omega_m - (1 - X) \log \Omega_1^0 - X \log \Omega_2^0,$$

where Ω_m is the configuration partition function of the mixture and Ω_1^0, Ω_2^0 respectively those for the pure substances 1 and 2. X is the atomic proportion of species 2.

We will apply these formulae to two relatively simple cases, in order to show the conclusions which may be drawn from this theory.

7.4. Ideal Solid Solution

As we have seen, an ideal solid solution is a solution in which the atoms, whatever they are, have properties which cannot be differentiated, but within which each atom retains its specific identity. This is what takes place with a mixture of white balls and black balls differing only in their colour, or to a first approximation with a mixture of isotopes of the same element.

We will find an expression for the partition function of such a solution.

Let there be a crystal containing a fixed lattice of atoms A and atoms B. The structure of the lattice determines, for each atom, a number z of immediate neighbours, known as nearest neighbours. This number z is equal to 6 for a simple cubic crystal, to 8 for a body-centred cubic crystal and to 12 for a compact stacking, face-centred cube or compact hexagonal lattice.

If we have N_A atoms forming the pure crystal C_A, we will have for these N_A atoms of A the total energy $-N_A E_{AA}$ taking into account that the energy is zero for the state corresponding to the atoms of A separated by an infinite distance.

By neglecting the interactions between atoms which are not nearest neighbours, we can say that the energy of interaction between two atoms of A in the pure crystal C_A is:

$$-2E_{AA}/z\,.$$

In fact an atom forming part of the crystal has $z/2$ bonds.

In the same way, the bonding energy between two neighbouring atoms of B in the pure solid B is $-2E_{BB}/z$. In the mixed crystal there will be pairs AA, BB and AB.

For the pairs AB it is convenient to write the bonding energy in the following form:

$$-(E_{AA}+E_{BB}-\omega)/z,$$

where ω is an exchange energy and when we put two pure crystals A and B in contact, the passage of an atom of A into B and an atom of B into A gives rise to an energy change 2ω.

When an AA pair is destroyed, a BB pair is automatically destroyed also and two AB pairs are formed.

By definition an ideal solution is a crystal for which $\omega=0$.

The bonding energy AB of two atoms A and B is then the arithmetic mean of the bonding energies of the pairs AA and BB.

Since $\omega=0$, the exchange does not lead to an increase in the internal energy. Thus *all configurations of atoms have the same energy.*

This enables us to calculate the partition function in this particularly simple case.

Let N be the number of atoms in the crystal. We will have NX atoms of B and $(1-X)N$ atoms of A, X being the atomic fraction of element B.

The energy for N atoms of the mixed crystal is then:

$$-N(1-X)\,E_{AA}-NXE_{BB},$$

whatever may be the distribution of the atoms in the lattice.

The number of arrangements of atoms A and B may be written:

$$\frac{N!}{[N(1-X)]!\,[XN]!}.$$

This gives for the configuration partition function:

$$\Omega = \frac{N!}{[N(1-X)]![NX!]} \exp\{[N(1-X)\,E_{AA} + \\ + NXE_{BB}]/kT\}.$$

In the summation the constant exponential term may be put as a factor for each energy state.

The free energy of configuration per atoms is given by:

$$G = -\,kT \log \Omega$$

$$G = -\,N(1-X)\,E_{AA} - NXE_{BB} - \\ -\,kT \log \frac{N!}{[(1-X)\,N]!\,[NX]!}.$$

Since the number of atoms N is large we can apply Stirling's theorem. Let us recall the demonstration of that formula:

$$\log N! = \log N + \log(N-1) + \cdots \log 2 \\ = \sum_{1}^{N} \log N,$$

and by assimilating the sum to an integral:

$$\sum_{1}^{N} \log N = \int_{0}^{N} \log N \, \mathrm{d}N = N \log N - N,$$

whence the approximate expression for G:

$$G = -N(1 - X)\,E_{AA} - NXE_{BB} + \\ + NkT\,[(1 - X)\log(1 - X) + X\log X\,.]$$

If we have two crystals, one containing N_A atoms of A, and the other containing N_B atoms of $_B$, the value of G for the two crystals together is:

$$G = -(1 - X)\,NE_{AA} - XNE_{BB}\,,$$

and by difference the change in free energy of mixture is:

$$\Delta G_m = NkT\,[(1 - X)\log(1 - X) + X\log X]\,,$$

i.e. for one gram-atom:

$$\boxed{\Delta G_m = RT\,[(1 - X)\log(1 - X) + X\log X]\,.}$$

We note that no restriction has been placed on the values of E_{AA} and E_{BB} and that these quantities may be different.

We no longer take into consideration the atoms which are not near neighbours. This implies that long-distance interactions are taken to be equal but not zero. This is a less restrictive hypothesis than regarding them as zero.

In the case of an ideal solution of two elements A and B, the energy of mixing is zero and the excess free energy on formation of the solution results only from the configurational entropy.

To take this into account we have only to calculate:

$$\Delta S_m = -(\partial G_m/\partial T)_P,$$

which may be written:

$$\Delta S_m = R[-(1-X)\log(1-X) - X\log X].$$

This term is always positive, since X is less than unity.

We obtain again the expression met with previously, but we have given it a meaning in terms of atomic structure.

7.5. Regular Solutions

In this section we will again obtain the expression for the regular solution, starting from the partition function, and we will give the atomic interpretation for this. We will retain the terminology which we have just used for the ideal solution.

In a crystal containing N atoms, the number of pairs of near neighbours is equal to a total of:

$$\tfrac{1}{2}Nz.$$

Let zN_{AB} be the number of AB pairs for a given configuration. This number varies with the degree of order. If the separation is total, this number is very small.

We have as before $N_A = N(1-X)$ atoms of A and $N_B = NX$ atoms of B.

Consider the A atoms. Since zN_A is the number of

neighbours of A atoms, the number of A atoms which are near neighbours of other A atoms is equal to:

$$z(N_A - N_{AB}),$$

and the number of AA pairs is:

$$\tfrac{1}{2}z(N_A - N_{AB});$$

the number of BB pairs is in the same way:

$$\tfrac{1}{2}z(N_B - N_{AB}),$$

so that the energy of configuration is:

$$E_c = -(N_A - N_{AB})\,E_{AA} + \\ + N_{AB}(-E_{AA} - E_{BB} + \omega) - (N_B - N_{AB})\,E_{BB}$$

$$\boxed{E_c = -N_A E_{AA} - N_B E_{BB} + N_{AB}\omega.}$$

The problem then consists of determination of equilibrium, that is the mean value of E_c equal to the internal energy in macroscopic thermodynamic theory:

$$\bar{E}_c = -N_A E_{AA} - N_B E_{BB} + \bar{N}_{AB}\omega.$$

To evaluate $\bar{N}_{AB}$ we use the hypothesis that the solid solution is disordered. This will lead to a limitation of the summation of the partition function to the terms which correspond to the most probable configurations.

It has been shown, in fact, than when the number of configurations is very large, the frequency curve is bell-shaped with a sharp peak. We can then be content with taking the maximum value as the value of the sum. This amounts to neglecting the contribution from the less probable complexions.

Let us evaluate, in these conditions, the number of average AB bonds: $z\bar{N}_{AB}$.

We had $\frac{1}{2}zN$ pairs.

The probability of finding an A atom on any given site is:

$$P_A = N_A/N.$$

The probability of finding a B atom on a neighbouring site to this A atom is:

$$P_B = N_B/N.$$

The probability of having an AB bond starting from an A atom is therefore:

$$P_{AB} = P_A P_B$$

and starting from a B atom:

$$P_{BA} = P_A P_B.$$

The mean number of AB pairs is therefore:

$$2P_A P_B(\tfrac{1}{2}zN).$$

Then:

$$\boxed{z\bar{N}_{AB} = zN_A N_B/N = zN_A N_B/(N_A + N_B).}$$

Let us now calculate the free energy G.

By putting E_c in the expression for $Q=\sum \exp(-E_c/kT)$, we find:

$$Q = \sum \exp(N_A E_{AA}/kT) \exp(N_B E_{BB}/kT) \times \\ \times \exp(-N_{AB}\omega/kT),$$

where the summation is taken over all configurations.

Now there are $N!/N_A!N_B!$ configurations.

Since

$$z\bar{N}_{AB} = zN_A N_B/N = \text{constant}$$

and

$$E_c = \bar{E} = \text{constant},$$

we can put $\exp(-E_c/kT)=\exp(-\bar{E}/kT)$ as a factor in Q, leading to the expression:

$$G = -N_A E_{AA} - N_B E_{BB} + kT \log \frac{N!}{N_A!N_B!} + \\ + \frac{N_A N_B}{N}\omega,$$

and for the excess of free energy per mole:

$$\Delta G_m = RT[(1-X)\log(1-X) + X \log X] + \\ + X(1-X)\mathcal{N}\omega,$$

where $\mathcal{N}$ is Avogadro's number.

The internal energy or enthalpy will be:

$$\Delta E_m \simeq H_m = X(1-X)\mathcal{N}\omega = X(1-X)A_0.$$

In this expression the configurational entropy is equal

to the entropy of mixing of the ideal solution. We thus obtain again the relationship which had been used to define a regular solution within the framework of classical thermodynamics.

We also obtain again one of the features of this solution, that these thermodynamic properties are symmetrical with respect to 0.5.

It is to be noted that in the preceding we have assumed that the solution is totally disordered. Depending on the sign of ω, however, there will be an average number of AB pairs which will be smaller or greater than $\bar{N}$ as calculated for the disordered solution, if there is order over small distances or local segregation.

PART II

STRUCTURE OF ALLOYS NOT IN THE EQUILIBRIUM STATE

CHAPTER 8

EVOLUTION OF STRUCTURE TOWARDS EQUILIBRIUM. ACTIVATION ENERGY

Consider a system which is able to change reversibly between a metastable state and a more stable state. Each of these states corresponds to a minimum value of the free energy, whilst to pass from one to the other, it must pass through an intermediate state where the free energy is higher. Let us consider Figure 25. This shows the change in free energy of the system when for example an atom changes place with another atom, following the chemical potential gradient. The parameter will in this case be used to indicate the position of the atom.

In order to pass from state I to state II, G must change by $\Delta G = G_{\mathrm{I}} - G_{\mathrm{II}}$. The term ΔG is negative since the system is moving towards equilibrium. The intermediate state is known as the activated state because an increase in free energy is necessary to reach and to pass through it.

The difference between the initial free energy G_{I} and the free energy G_A of the activated state is known as the free energy of activation for the movement of the atom. The reaction can therefore proceed by a succession of such jumps as lead to a progressive reduction in the free energy of the system.

The energy necessary for reaching the activated state

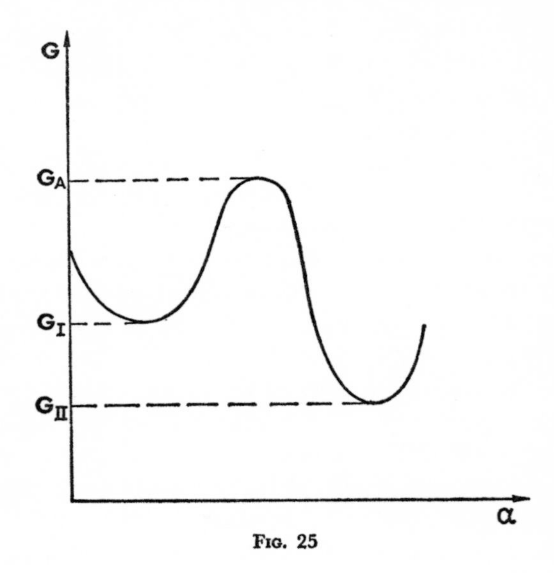

FIG. 25

may be supplied by thermal agitation. Each atom in fact vibrates about a mean position and the corresponding vibrations have a very wide energy spectrum.

For a change to take place in the system, the atoms must exchange places. It is only those atoms which acquire an energy greater than a certain activation energy which can move away. The speed of the reaction corresponding to this atomic movement therefore depends directly on the number of atoms which, at the moment in question, have sufficient energy. To find this number, we will calculate with the aid of statistical thermodynamics the number of atoms which, at each instant, are in a specified energy state; we will then find the sum for all energy states greater than the activation threshold.

The energy levels which the system can occupy or which an atom can occupy are discrete. We will designate then by $e_1, e_2, \ldots, e_N$ for the N atoms.

If certain energy states have the same level, and if for a level i the number of these is g_i, the proportion of atoms occupying state e_i is equal to:

$$\frac{n_i}{N} = \frac{g_i \exp(-e_i/kT)}{\sum_1^N g_i \exp(-e_i/kT)}.$$

This distribution is known as Maxwell-Boltzmann, or simply Boltzmann. It applies in the case of atoms since, due to their mass and velocity, they obey the laws of classical dynamics.

Let us calculate the number of atoms which have at time t an energy equal to or greater than the energy of the activation barrier, U_A per atom. For this purpose we must know the distribution of energy.

In the case of a crystal, the thermal energy is due to the vibrations of the atoms about their positions. Einstein suggested a crystal model consisting of atoms vibrating independently of each other. This leads to assuming the existence of $3N$ modes of vibration. In these conditions, we have for frequency ν:

$$e_i = (i + \tfrac{1}{2})\, h\nu/kT\,,$$

where here i is the quantum number of level i and h is Planck's constant. This model, therefore, supposes that the frequencies are equally spaced. From this we find the partition function Q:

$$Q = \sum_0^\infty \exp\left[-\,(i + \tfrac{1}{2})\, h\nu/kT\right].$$

U_A is equal to $nh\nu$, if n corresponds to the level U_A.

The proportion of atoms at level U_A is:

$$Q^{-1} \exp\left[-\,(n + \tfrac{1}{2})\, h\nu/kT\right].$$

A similar calculation gives the proportion of atoms with energy greater than U_A, and for the total we have:

$$\begin{aligned} f &= Q^{-1} \exp\left[-\,(n + \tfrac{1}{2})\, h\nu/kT\right] \sum_{i=0}^{i=\infty} \exp(-\, ih\nu/kT) \\ &= \exp(-\, nh\nu/kT) \\ &= \exp(-\, U_A/kT)\,. \end{aligned}$$

The distribution model is too simple to be generally true, but it provides an approximation which suffices for the end we seek. We can state the following theorem:

In any system in equilibrium comprising N atoms, at constant volume and temperature, the proportion of atoms possessing an energy greater than a specified value U_A is proportional to

$$\exp(-\, U_A/kT)\,.$$

Let us apply this theory to the case of a reaction.

If the reaction takes place by means of one single elementary process, and if the activation energy of this process is independent of temperature, the velocity $\mathrm{d}y/\mathrm{d}t$ at which the system is transformed (y=part transformed) is:

$$\mathrm{d}y/\mathrm{d}t = pv \exp(-U_A/kT).$$

The reaction rate is in fact proportional to the fraction of atoms with an energy level greater than U_A and to the probability, during the time when the energy is greater than U_A, that they have of satisfying the geometrical conditions which permit a local transformation.

p is really the ratio of the number of complexions associated with the intermediate state to the number of complexions associated with the initial state, i.e.:

$$p = W_A/W_I,$$

whence, according to Boltzmann:

$$p = \exp(S_A/k),$$

and

$$\begin{aligned}\mathrm{d}y/\mathrm{d}t &= v \exp(S_A/k) \exp(-U_A/kT)\\ &= v \exp(-\Delta G_A/kT),\end{aligned}$$

where ΔG_A is the term for free energy of activation.

In general, Arrhenius' formula is used:

$$\mathrm{d}y/\mathrm{d}t = A \exp(-\Delta G/kT).$$

The velocity of a reaction depends therefore to a large

extent on the value of U_A. The higher the energy of activation, the greater the need to work at high temperatures in order to attain an appreciable change in the system.

Table I gives numerical examples of Arrhenius' factor for different values of ΔG and for different temperatures. We see that the velocity depends greatly on the ratio $\Delta G/T$.

When the reaction no longer takes place in the form of a single elementary process, the expression for the activation energy is not so simple. If, however, two processes are involved for the reaction to take place, one may be much easier than the other.

We shall see in what follows that the rate of precipitation of a phase in a supersaturated solution depends on the rate of formation of nuclei and on the rate of migration of the atoms which contribute to its growth. Two extreme cases are usually found. Firstly, nucleation is very rapid and diffusion slow, and this latter process controls the change. Alternatively, nucleation is very slow, at high temperatures, for example, diffusion is then rapid, and the nucleation process controls the progress of the precipitation.

TABLE I

Values of $\exp(-\Delta G/kT)$ for different temperatures

ΔG kcal/at. g	Temperature (K)		
	300	1000	1500
40	0.718×10^{-29}	0.181×10^{-8}	0.148×10^{-5}
50	0.372×10^{-36}	0.118×10^{-10}	0.518×10^{-7}
60	$< 10^{-38}$	0.768×10^{-13}	0.181×10^{-8}

CHAPTER 9

HARDENING BY PRECIPITATION

9.1. Mechanical Properties

We said in the introduction that metallic alloys have mechanical properties which are far more interesting than those of pure metals. By means of alloys we are enabled to manufacture machine parts which are lighter, or more resistant for the same weight, than those made from pure metals.

These mechanical properties are: resistance to deformation and ductility.

The resistance to deformation may be determined by subjecting a sample to a load, in tension or in compression, and measuring the force at which a specified elastic deformation, then a permanent deformation – elastic limit – are obtained. We then calculate the stress, force per unit surface area, which does not depend on the dimensions of the sample.

The elastic limit is not always well marked, so we more usually determine the conventional elastic limit which corresponds to an elongation of 0.2%. Finally, we use the ultimate tensile stress, which is the maximum value reached by the stress during a tension test. The conventional yield strength and the ultimate tensile stress are vital characteristics in calculation of the dimensions and shapes of components subjected to loads during the operation of machines.

The ductility is the readiness with which a material acquires a permanent deformation. It is measured by determining the deformation corresponding to maximum stress. An ideal material must possess both a high elastic limit and a high ductility. If, in fact, the stress encountered during operation should locally reach a value greater than the elastic limit, a small plastic deformation is produced together with a relaxation of the stress. If no plastic deformation is possible, fracture is immediate and often catastrophic.

Ductility is a property necessary for the cold forming process.

9.2. Hardness

The mechanical resistance may be characterised just by a measurement of the hardness. This measurement consists of a determination of the depth to which a indentor enters under the action of a force on a polished face of a sample.

The indentor is a ball or a point cut into the shape of a pyramid, and we measure the surface of the part of the metal which has been depressed by the pressure applied.

For a hard sample, the indentation will be of small area, whilst it will be of large area for a soft sample.

The hardness test is easy to carry out, since it only

necessitates small samples with only one surface polished. For tensile testing, on the contrary, we use samples of relatively complicated shape. One usually finds an empirical relationship between hardness and tensile resistance. A measurement of the hardness therefore allows us to determine the approximate resistance of an alloy and to select a limited set of conditions for carrying out tensile tests which give not only the tensile strength but also the ductility.

9.3. Influence of the Microstructure on the Resistance to Deformation

Certain alloys are more resistant or more plastic than others because of their particular microstructures.

As we said at the start of this work, metals and alloys are formed of small crystals fixed together.These crystals are not perfect. They contain defects such as dislocations, atoms of impurities or of alloy, and dispersed particles.

Dislocations are linear defects of which the displacement leads to plastic deformation. Indeed, a crystal can only be subjected to deformations without modification to the crystallographic structure by sliding along a crystallographic plane.

Such a sliding could take place, as shown in Figure 26, by displacement en bloc of the two parts of the crystal situated on the two sides of the slip plane. From this there would result the formation of two steps of height *b*.

This mechanism necessitates enormous shearing forces.

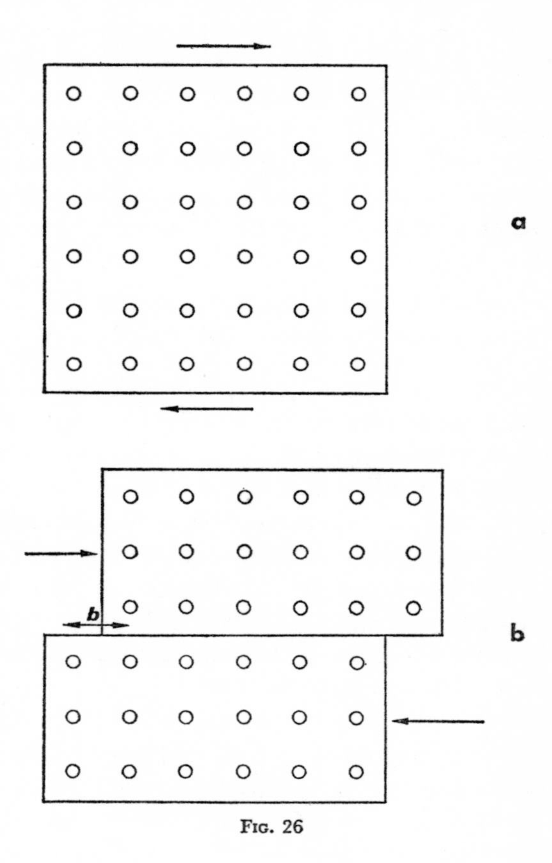

FIG. 26

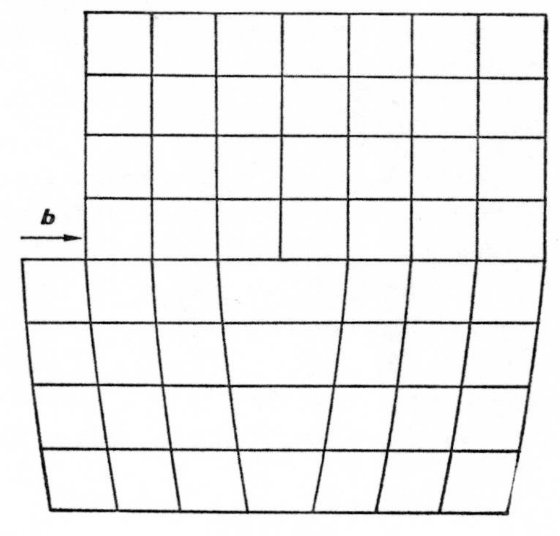

FIG. 27

On the other hand, the same deformation is easily obtained by passage of a dislocation from one side to the other side of the crystal. Dislocation results in the formation of one single step of height b parallel to the slip plane. This is illustrated in Figure 27, which shows schematically the structure of an edge dislocation.

There is another type of dislocation, the screw dislocation, which also produces by its displacement a slip-type deformation.

A substantial macroscopic deformation is caused by the formation and subsequent displacement of a large number of edge and screw dislocations. The displacement may be hindered by obstacles such as grain boundaries, atoms in solution or dispersed particles.

Effects of grain boundaries

The deformation of a crystal is possible only if the dislocations reach the external surface to form a step. An internal stress is produced if an obstacle blocks them inside the crystal. The force necessary to produce the deformation is larger than in the absence of obstacles.

Since grain boundaries form very effective obstacles to dislocations, it follows that the resistance to deformation of alloys or metals increases in inverse ratio to the grain size.

Effects of elements in solid solution

The presence of foreign atoms in solution in a crystal causes local changes in bonding energy between neighbours and in the distance between units of the lattice. From this results a hindrance to displacement of the dislocations, and hence an increase in mechanical resistance. This increase is small in magnitude.

Effects of dispersed particles

The presence of very fine particles distributed within crystals leads to a considerable increase in resistance to deformation. This is the most controllable and the most effective means of hardening of alloys. Its application is increasing rapidly.

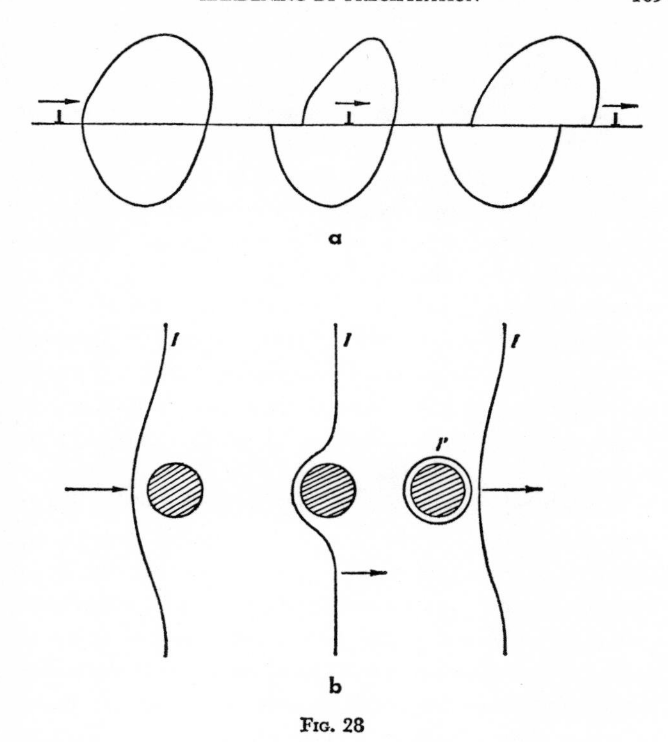

FIG. 28

The principle of this hardening is as follows: we form within the alloy a very large number of very small crystalline domains, with properties different from those of the matrix. These domains form very effective obstacles to movement of dislocations if their own resistance to shear is very large. They are sheared or passed over by dislocations, as shown in Figure 28.

They are sheared when they are very small. It is then necessary to apply a supplementary force to overcome the increase in internal cohesion at the place in question, and to supply the energy necessary for creation of a supplementary chemical interface at the steps created by passage of the dislocation (Figure 28a).

Large domains, small in number, are passed over. The dislocations move over them, leaving a loop in their slip plane (Figure 28b) or merely changing the slip plane. In either case the length of the dislocation line is increased, which needs a greater force to displace it through the crystal.

Theoretical calculations and experimental results (observations by electron microscope) show that numerous small precipitated domains are more effective, for equal volumes, for hardening than loosely packed large domains.

We will see that decomposition by tempering or by ageing of supersaturated solid solutions leads to the formation of numerous fine precipitates which cause an increase in mechanical resistance. This precipitation is possible only in certain alloys.

9.4. Conditions to be Satisfied by Alloys for Hardening by Precipitation

Let us consider the simple case of a binary system forming

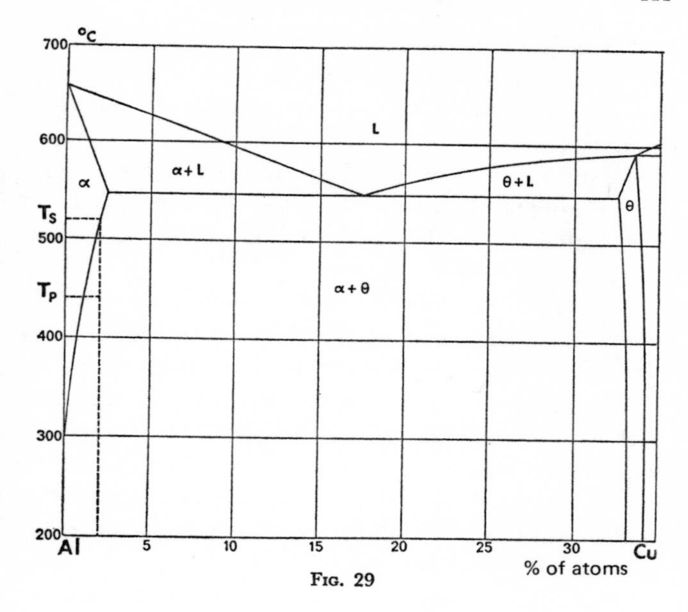

FIG. 29

a terminal solid solution the extent of which increases with temperature. The aluminium–copper system is a good example of this (Figure 29).

The alloy of composition X is in a state of a homogeneous solid solution above T_S, the solubility limit of the compound θ (Al_2 Cu). The atoms of solute (copper) are then distributed at random over the sites in the solvent. On the other hand, below T_S the alloy is constituted from two phases in equilibrium: here one is the compound θ, but in an other system it could be a terminal solid solution; the other is the solvent which contains fewer copper atoms in solution as the temperature falls, at constant concentration.

Hardening takes place when the alloy is heated to a temperature slightly above T_S, then suddenly placed in a temperature T_P less than T_S, or quenched to room temperature and reheated to temperature T_P.

We find, in general, that hardening increases with the degree of supersaturation, that is the quantity of solute which can precipitate, and the more dispersed the phase.

It is, therefore, advantageous to select a composition such that at temperature T_P the material is well inside the two-phase domain so as to have a relatively large volume of precipitated phase, and to carry out the precipitation at the lowest possible temperature so as to avoid growth of the precipitated material.

In practice we are limited by the size of the domain where the terminal solid solution exists, or by the homogenisation temperature. If we wish to have the greatest degree of supersaturation we must work at the concentration corresponding to the highest temperature T_S. In many cases T_S is then such that oxidation may be produced (in the region of 1100° C for steels) with serious consequences for the properties of the alloy (loss of metal, change in composition by selective oxidation of certain elements).

A good system will have a limiting solubility curve with

a small slope allowing substantial supersaturation after homogenisation at an acceptable temperature.

The ageing temperature should not always be low, since this would lead to too long a treatment time; this time must not, in practice, exceed about ten hours. Precipitation is a function of the mobility of the atoms in the metal, and this mobility follows Arrhenius' law in its dependence on temperature. In the case of austenitic and ferritic steels, ageing treatments are generally carried out above 550° C. Martensitic alloys are usually treated between 400° C and 500° C. Ageing of light alloys or copper-based alloys takes place between 20° C and 300° C.

There exist, at the present time, a large number of binary alloys which can be hardened and which have practical applications. Amongst these there are aluminium-based alloys hardened by copper (already mentioned), nickel-based alloys hardened by titanium and aluminium, copper-based alloys hardened by beryllium, and finally, the latest developments which are nevertheless very promising, carbon-free iron alloys hardened by titanium, molybdenum, niobium, etc.

It appears that the most interesting results are obtained, as in the case of steels, when several elements are added together. Thus, iron is hardened considerably by nickel and titanium, or by titanium and silicon, or by nickel and manganese. The presence of these two latter elements alone does not produce hardening.

9.5. Change in Hardness as a Function of Ageing Time

We will now give some examples of hardening curves as functions of time and temperature.

Figure 30 shows the hardening curves obtained with an aluminium alloy containing 4% of copper, aged at different temperatures.

The alloy was first brought to 520° C for several hours, then quenched. It is then homogeneous and in the supersaturated state. All the copper atoms are in solid solution

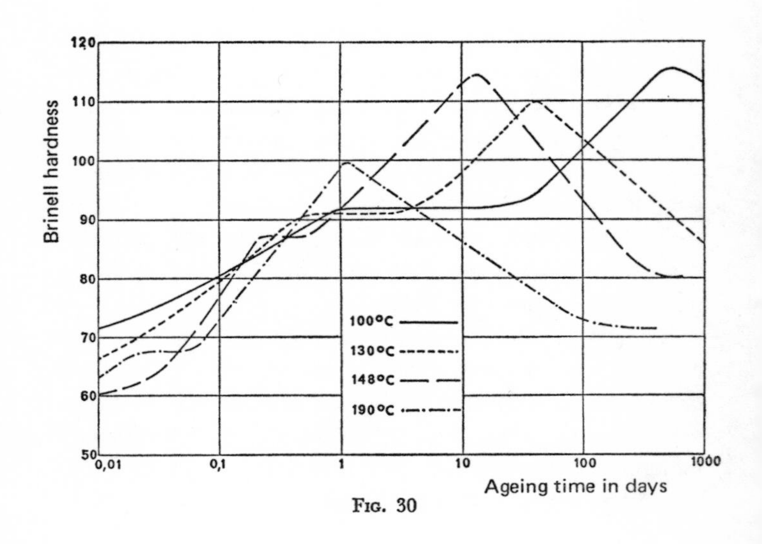

Fig. 30

in the aluminium, since, during quenching, they do not have the time to travel long distances.

The alloy was then subjected to ageing treatments at constant temperatures, for increasing times during which the hardness was measured.

We see from Figure 30 that the hardness increases as a function of the ageing time, that the rate of hardening depends on the temperature, and that the shape of the curves also depends on the temperature. In particular, the maximum value is lower at high temperatures than at low temperatures.

Figure 31 shows the hardening curves obtained with an alloy of Fe–Ni–Ta. The homogenisation treatment was carried out at a very high temperature, 1200° C.

The face-centred cubic structure which is in equilibrium at that temperature, is preserved by quenching. Ageing must take place above 500° C if the changes in hardness are to be measurable after tempering times less than 1000 hours.

We also see here that the rate, the maximum value of the hardness and the general form of the curves are greatly dependent on the temperature of ageing. We note in particular that the time required for the onset of hardening is greater at high temperatures, despite the greater mobility of the atoms.

Figure 31 also contains a curve obtained from samples which were cold-worked after quenching. The presence of dislocations left by the deformation of a plastic nature

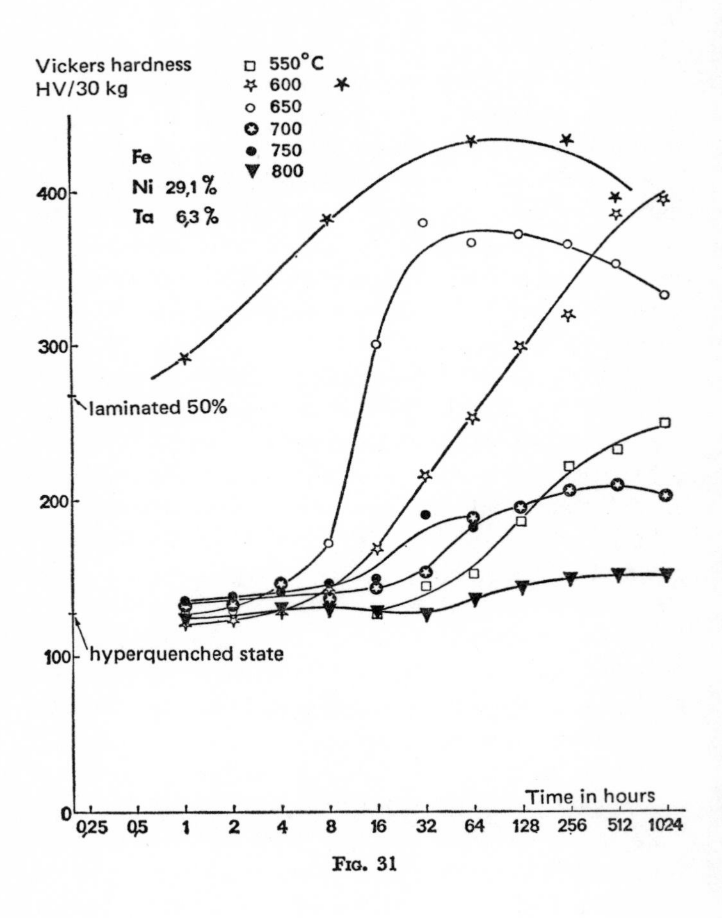
Vickers hardness
HV/30 kg
550°C
600
650
700
750
800
Fe
Ni 29,1 %
Ta 6,3 %
laminated 50%
hyperquenched state
Time in hours
0
100
200
300
400
0,25 0,5 1 2 4 8 16 32 64 128 256 512 1024

FIG. 31

leads to a far greater rate of ageing than when there was no cold-working.

These same phenomena are seen in the tensile properties. The resistance to tensile stress increases, reaches a maximum value, then decreases again. The ductility falls as the resistance increases. The elongation varies from 40% in the quenched state to less than 10% when the resistance is at a maximum.

If we compare the results of tensile tests with the results obtained from hardness measurements we naturally find a good correlation. The resistance to tension increases along with the hardness. On the other hand, the elongation decreases when the other two quantities increase.

When the ageing time is very long, especially at high temperatures, the values of the mechanical properties again become very close to those observed in the quenched state. We will describe the alloy as overaged or softened. This same result is found if, in place of quenching the alloy, it is allowed to cool slowly starting from the homogenisation temperature.

9.6. Changes in Crystallographic Structure During Hardening

Hardening does not in general correspond to formation of the equilibrium compound with a structure differing from that of the matrix. After quenching the atoms of solvent and solute are distributed at random over the sites in the

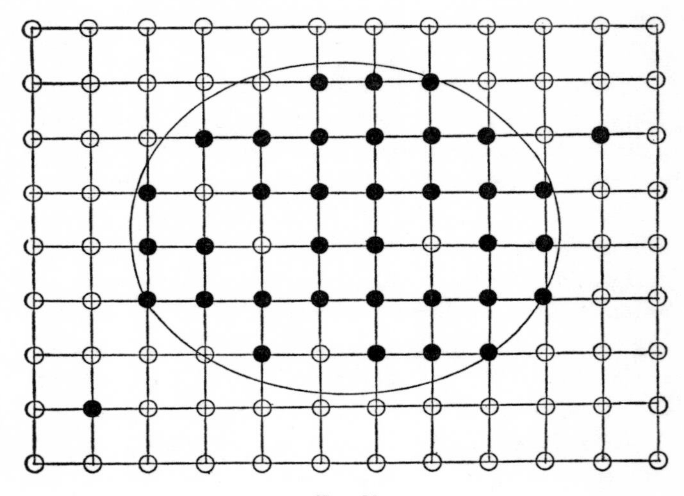

Fig. 32

crystal. As soon as the ageing temperature is reached, the atoms of solute tend to occupy new positions so as to reduce the free energy of the alloy. They diffuse, following the chemical potential gradient, forming domains where their concentration is larger than in the untransformed matrix.

In these domains, the sites occupied by the atoms are at positions only slightly different from those of the sites in the initial crystal. These are domains which are coherent with the matrix. The atomic planes traverse them without discontinuity, despite the local change in chemical composition. Clearly, the atoms of solute always have a different atomic number from those of the solvent, but they may be either of the same size, larger, or smaller. When they are of the same size, the domain is perfectly coherent and its existance is shown only by a local change in composition (Figure 32).

If the atoms of solute and solvent are of different sizes, the atomic sites are displaced with respect to the positions in the matrix, as is shown by Figure 33, without, however, introducing any discontinuity.

Such displacements lead to a local increase or decrease in the lattice separation.

Finally, if the two elements tend to form ordered compounds, the atoms take up positions inside the domain at privileged sites in order to form a superstructure (Figure 34). This may or may not result in a local change in lattice separation.

To sum up, many sorts of coherent domains may appear:

(a) disordered domains without displacement of atoms;
(b) ordered domains without displacement of atoms;
(c) disordered domains with displacement of atoms;
(d) ordered domains with displacement of atoms.

In addition to these differences in internal structure, the domains may assume varied external shapes. They may be spherical, in the shape of small cubes, in the shape of small plates, and more rarely in the shape of rows of atoms.

Formation of coherent domains, with structure not

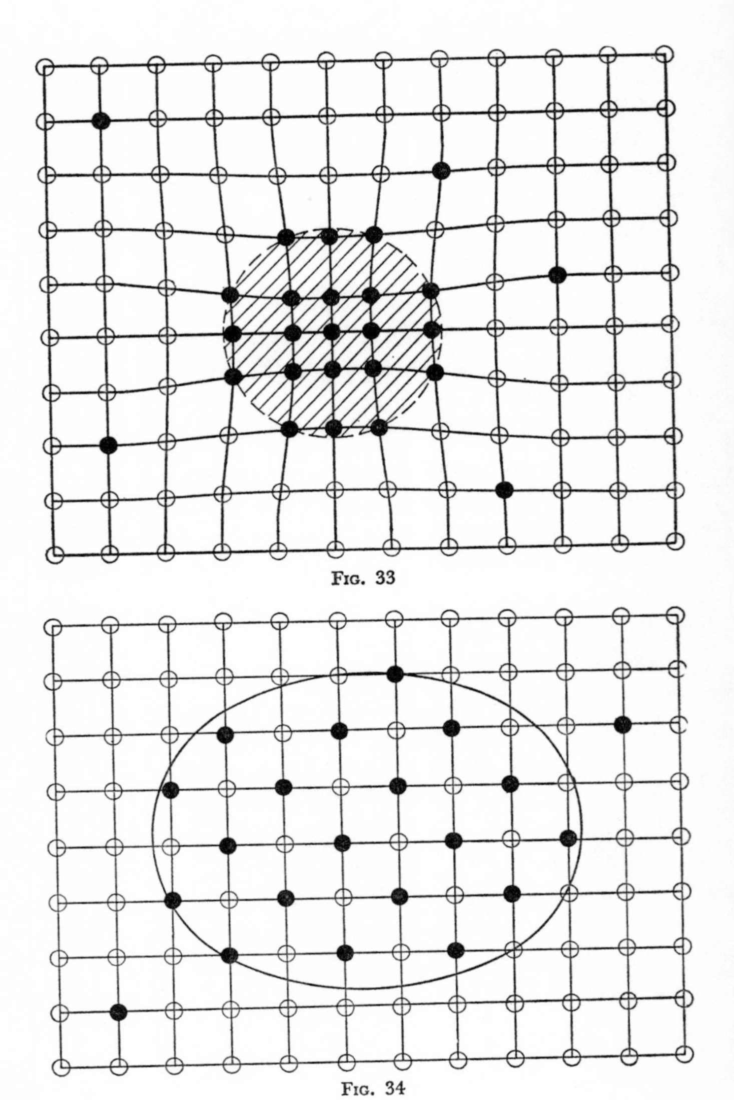

Fig. 33

Fig. 34

greatly different from that of the matrix, is generally the first step of an evolution towards a metastable state of equilibrium, since the equilibrium precipitate is almost always very different in structure from the matrix. Certain alloys have several metastable equilibrium states to which there correspond different metastable precipitated phases.

As illustrations of this aspect we will consider two examples, one in the aluminium-copper system and the other in the nickel-titanium system.

Alloy Al–4% Cu

As we have said above, the alloy Al–4% Cu is the type of alloy hardened by precipitation. Its hardening was discovered by Wilm in 1905. Merica later explained this by the rejection of copper atoms from the supersaturated solid solution. As is shown by Figure 29, equilibrium corresponds to the coexistence of a solid solution poor in copper and of the compound θ of formula $CuAl_2$. Precipitation of this phase is not direct except at very high temperatures. At low temperatures, where the greatest hardening effect is observed, it does not have the time to form. We find, however, successive changes in structure. After quenching the solid solution decomposes as soon as room temperature is reached. The copper atoms assemble to form plane zones rich in copper as shown in diagram form in Figure 35.

The centre of each zone is constituted of a single plane of copper atoms parallel to the $\{100\}$ planes of the lattice.

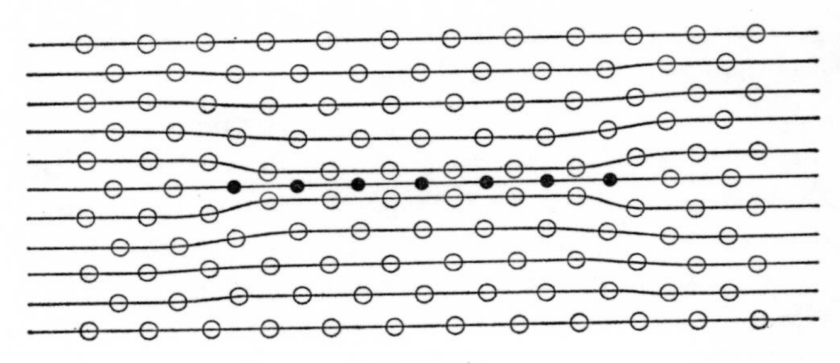

FIG. 35

These atoms are smaller in size than the aluminium atoms. This results in a reduction in the lattice spacing in the very centre of the zone. The distance between lattice planes is therefore larger in those parts of the lattice in the immediate neighbourhood of the zone, so as to ensure coherence. Such assemblies of copper atoms are called Guinier–Preston zones, from the names of the two researchers, who independently established their existence by means of X-ray diffraction in 1938.

When the ageing time is sufficiently long, the formation of another metastable phase is detected, the appearance of which coincides with the disappearance of the Guinier–Preston zones as they enter into solution. This phase is known as θ'' by Guinier and his followers. It corresponds to discs 25 Å thick and about 150 Å in diameter within which the atoms are ordered. Its composition differs from that of the G–P zones. There the concentration in copper is lower. The discs of phase θ'' also appear coherent with the lattice and are parallel to the {100} planes. G–P zones are only observed when the ageing takes place below 200° C. Only θ'' causes hardening in the region of 200° C.

Above 200° C we observe another metastable compound, either after formation of θ'' or directly, of which the structure is very different from that of the matrix. This is the tetragonal compound θ' which is seen in the form of fine visible discs under the optical microscope. This phase is not perfectly coherent with the matrix; it subsequently disappears when the equilibrium compound Al_2Cu is precipitated. This latter compound itself has in general a globular form and is totally incoherent with the matrix.

To sum up, decomposition of the alloy takes place, slightly below 200° C, in accordance with the successive reactions:

solid solution $\rightarrow$ G–P zones + matrix 1 $\rightarrow \theta''$ + matrix 2 $\rightarrow \theta'$ + matrix 3 $\rightarrow \theta$ + matrix 4.

Matrices 1, 2 and 3 coexist for a short time with the different metastable phases. They differ in their composition; their copper content decreases with their stability.

Figure 36 shows the limits of the domains of temperature and time within which the different structures are observed, for differing initial concentrations of the Al–Cu alloy.

Electron microscope observations have been made in recent years by transmission through thin foils. They show that during hardening the alloy undergoes changes in structure on a very fine scale, but it is extremely difficult to

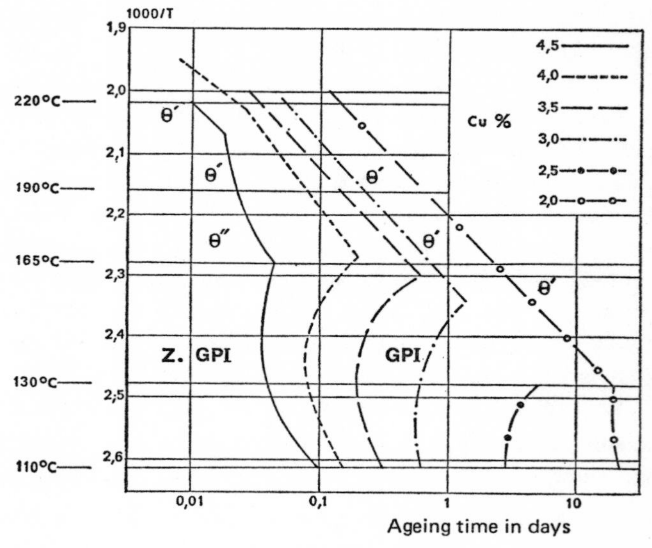

FIG. 36

distinguish the details. Interpretation of these pictures was made possible because X-ray diffraction had shown the existence of such zones. X-ray diffraction is a technique which gives information about perturbations of the lattice on the scale of the unit cell. Formation of G–P zones in the solid solution causes local perturbations in the periodicity of the lattice and the X-rays are no longer diffracted in a simple manner. The intensity of reflections from the matrix is altered and abnormal diffusion appears. From the nature of this latter, we obtain information concerning the mean dimensions, thickness and diameters, and the internal structure of the G–P zones.

Nickel alloy containing 9% titanium

Alloys of Ni and Ti with titanium content between 6% and 9% by weight are hardened by ageing between 500° C and 800° C. Titanium is soluble in nickel in the solid state up to about 12% approximately, at 1300° C. Solubility decreases with temperature and falls to 6%, at 500° C. The limit of solubility at 1100° C is about 10%.

The homogeneous solid solution is face-centred cubic. It is in equlibrium with the Ni_3Ti phase, the structure of which is hexagonal. Direct precipitation of this latter phase is only produced above 800° C. Below this temperature the formation of a coherent ordered compound is observed, but firstly there takes place a process which is quite special. The composition changes locally and regions alternatively rich in nickel or in titanium appear, in three perpendicular directions corresponding to the ⟨100⟩ axes of the f.c.c. crystal of the matrix.

The change in composition along one of the ⟨100⟩ directions is shown diagrammatically on Figure 37. The distance between two neighbouring titanium-rich domains is appreciably constant throughout the crystal. The structure is modulated in this manner.

Modulation is quasi-periodic over a distance of several thousand angstroms; the periodicity depends on the tem-

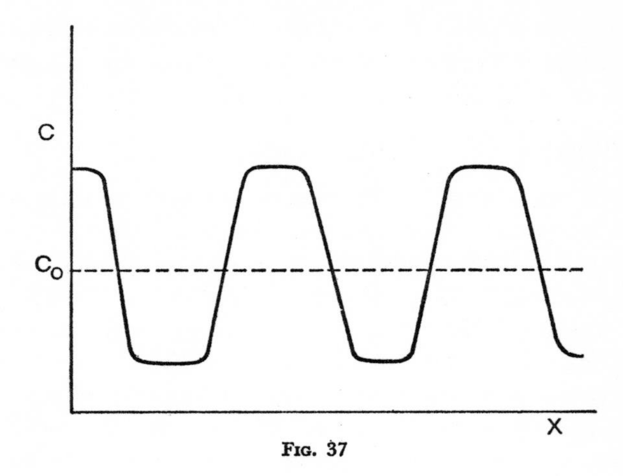

Fig. 37

perature and time of the ageing. It lies between 50 and 1000 Å. Hence the X-ray diffraction patterns have a very characteristic appearance. The diffracted lines corresponding to the supersaturated lattice do not appreciably change in position; they are accompanied by supplementary lines, called side bands, the position of which is connected with the period of modulation. In this case the coherent domains are of relatively large extent, in contrast to the Guinier-Preston zones which are very small. It appears that we can classify alloys, corresponding to these two types of coherent domains, in two categories: alloys with G–P zones and alloys with modulated structure. The only common feature is the existence of several stages of decomposition: pre-precipitation, formation of metastable compound(s) and precipitation of the equilibrium phase.

9.7. Changes in Microstructure

The changes in crystallographic structure that we have just discussed are accompanied by changes in microstructure. The microstructure is the structure observed by means of the electronic microscope or by the optical microscope.

When they are large enough, it is possible to distinguish coherent or incoherent precipitates with the aid of the electron microscope or the optical microscope.

It is evident that observation of this nature is easiest in the overaged state. The precipitated domains are then large, of the order of 10μ, and their composition is very different from that of the surrounding matrix. Due to their structure, they always have a smaller chemical activity than the matrix. They can easily be distinguished from the latter by a selective chemical attack.

For observation with the optical microscope, the main sample is cut in such a manner as to expose a plane surface, which is polished by successively finer abrasives, and is then polished electrolytically. Specific reagents are used for each phase.

With the help of the electron microscope it is possible to distinguish precipitated domains of the order of 100 Å, when, for example, the hardness has reached its maximum value.

Two techniques are used for observation of the samples:

(1) a microrelief replica of the polished and etched surface of an aged sample is prepared and then examined by transmission;

(2) a thin piece of alloy in the aged state is thinned locally to a thickness of the order of 1000 Å and is examined directly by transmission.

The electron microscope shows that for a given temperature and time of ageing, the size of the precipitated matter varies slightly about a mean value, and that this mean value increases with the time taken, but that the volume remains in effect constant. This corresponds to the phenomenon of coalescence. In fact, small precipitates are unstable in presence of large, and they dissolve and help to increase the size of the latter. The total free energy of the alloy is the sum of the free chemical energy, proportional to the volume of each phase and therefore constant, and of the surface energy, proportional to the total surface of the precipitates.

For a given volume of precipitates the total interface area is smaller for larger precipitates. Hence the preferential dissolution of the small precipitates produces a reduction in the free energy of the alloy.

9.8. Widmanstäten Structure

The precipitates nearly always have geometrically simple shapes: spheres, cubes, discs, or needles. The edges of the

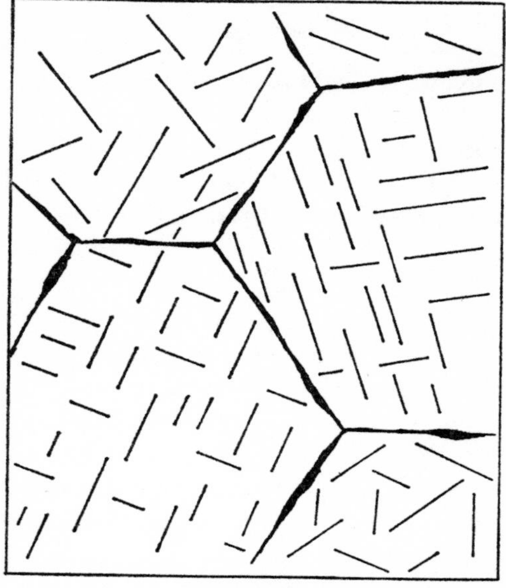

Fig. 38

cubes, discs or needles have orientations which are well defined with regard to the lattice of the matrix. In a given crystal we find discs or needles, which form families with the same orientation. The micrographic picture resulting from this is known as the Widmanstäten structure and was observed for the first time in a meteorite.

Figure 38 shows such a structure. The precipitates are

discs which are grouped in three families per crystal. In certain crystals only two families are visible, the discs of the third family being parallel to the observation plane.

Observation under the microscope shows also that precipitation takes place not only at the interior of the grains, but also at the boundaries, which then become thicker than in the alloy in the quenched state.

CHAPTER 10

THERMODYNAMIC CONSIDERATIONS ON FORMATION OF METASTABLE PHASES DURING AGEING OF VERY HARD ALLOYS

The equilibrium diagrams and the curves of free energy versus concentration corresponding to these do not predict the existence of the metastable phases which appear during ageing at low temperatures. It is therefore necessary to complement them by metastability diagrams.

As an example, let us take a binary alloy composed of elements A and B and where the precipitation of the equilibrium phase (stage III) is preceeded by a pre-precipitation stage (stage I) and by a stage during which there appears a metastable phase coherent with the matrix (stage II).

We have seen that it is possible to have, for a certain region of composition and temperature, stage III alone, stages II and III, or the three stages in succession. We can therefore, in theory, draw on the equilibrium diagram the curves showing the boundaries of the regions inside which each of the stages is observed (Figure 39).

The curves indicated by broken lines are the boundaries between the two-phase domains and the corresponding matrix which has been made to give up some of its constituents. It is only in exceptional circumstances that we can determine the position of the region of existence of metastable phases starting from element B.

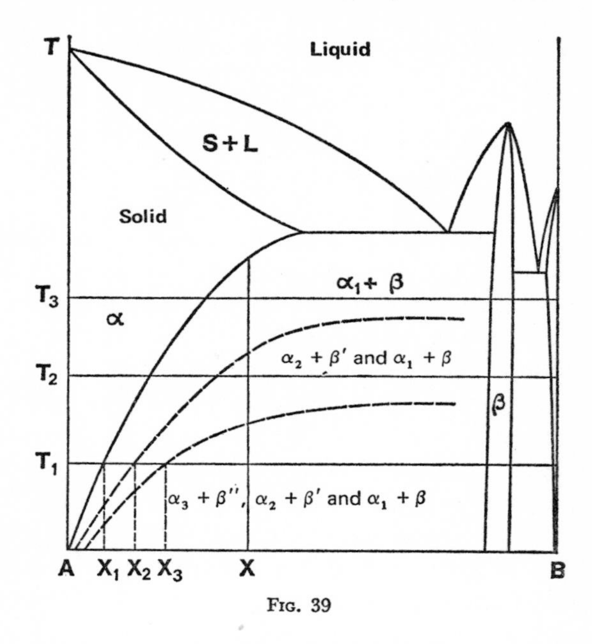

Fig. 39

In order to understand better what takes place, consider several temperatures: T_1, T_2 and T_3. At temperature T_1 the isothermal passes through the three precipitation regions. The form of the free-energy curve, if it could be experimentally determined, is shown in outline on Figure 40, where f is the free energy per atom of alloy. The full curves correspond to the equilibrium diagram. The curve for the terminal solid solution, in element A side, has been prolonged as a broken line and we have assumed that it

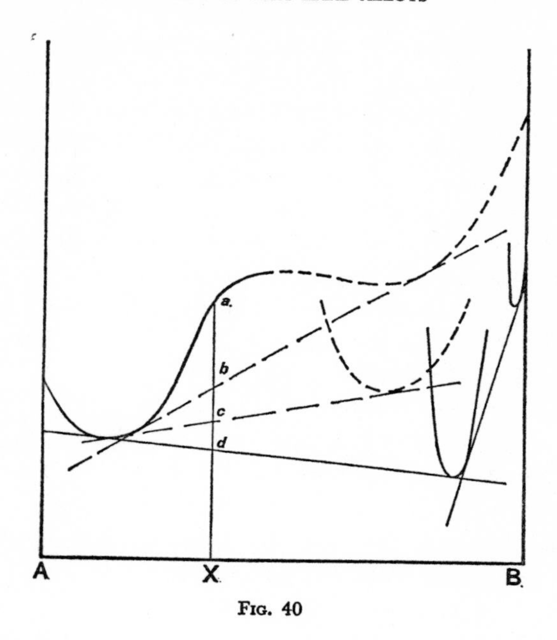

FIG. 40

would have a second minimum on the element B side, so that two solid solutions in metastable equilibrium can appear for composition X, for example. This solid solution rich in B will be called β''. It might correspond to an unstable structure of B, isomorphous of A, and appears for compositions beyond the point X_3 given by the tangent which is common to the two minima.

This figure also shows, as a dotted line, a free-energy

curve for a hypothetical metastable intermetallic compound. Its minimum is higher than that of the minimum for the free-energy curve for the equilibrium intermetallic compound. This metastable compound will appear above composition X_2 also corresponding to the common tangent with the curve for the solid solution of B in A. This compound, which we call β', will be more stable than β'', but unstable in presence of β. (The relative position of the compositions corresponding to the minima on Figure 40 is arbitrary.) The compositions X_1, X_2 and X_3 are the same as on the diagram in Figure 39.

If we consider temperature T_3, the free energy diagram would be as outlined in Figure 41. In this case the free energy curve for the solid solution no longer has more than one minimum and the curve corresponding to the metastable intermetallic compound lies higher up. Only the equilibrium precipitate can appear.

At the intermediate temperature T_2 we would have the curve for the metastable compound lying between the curve for the equilibrium compounds and the curve for the solid solution; this latter would have only one minimum. The isothermal at T_2 passes through only two regions in Figure 39.

In the special case of the alloy Al–4% Cu, it is necessary to imagine a more complicated free energy diagram, explaining the appearance of the four stages which correspond to the zones and to the phases θ'', θ', and θ. We would then find, between the curves corresponding to θ'

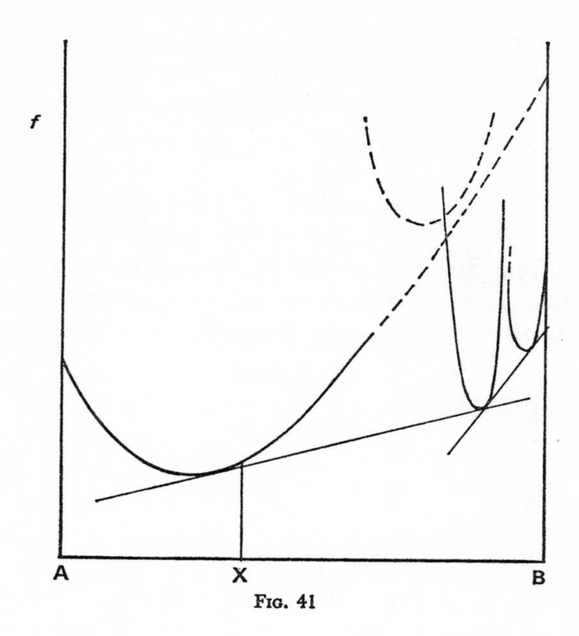

FIG. 41

and θ of our theoretical example, a further curve for the metastable precipitate θ''.

Why does the stable phase not precipitate directly?

Since its formation corresponds to the greatest change in free energy starting from the supersaturated solid solution, the driving force leading to its precipitation is the greatest. We see, in fact, from Figure 41 that the change in free energy for each stage goes in increasing order:

$\Delta f_1 = a - b$ for pre-precipitation;
$\Delta f_2 = a - c$ for the metastable compound;
$\Delta f_3 = a - d$ for the equilibrium phase;

$$\Delta f_1 < \Delta f_2 < \Delta f_3 .$$

The kinetics are, however, not controlled by the magnitude of the change in free energy. We will see later that the precipitation of phases by nucleation or by other mechanisms necessitates an energy of activation. If a system in a metastable state is able to pass into other metast-

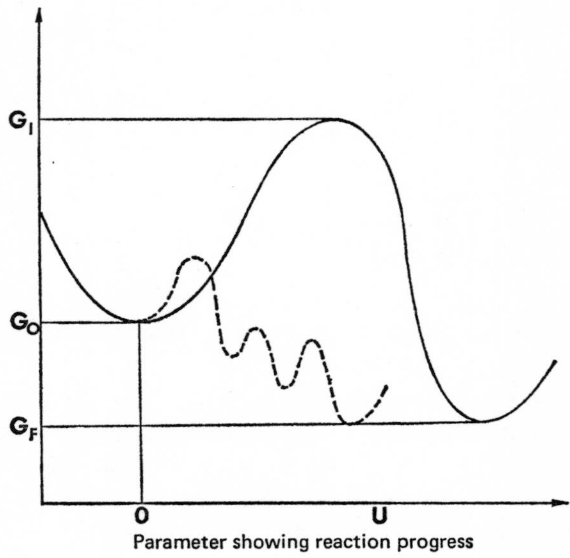

FIG. 42

able or stable states, it will be more able to pass immediately into the state for which it will need to acquire the smallest energy of activation. The system evolves by following the path which calls successively for the smallest activition energies. Experiment shows that it is easier at low temperatures to pass from solid supersaturated solution to the pre-precipitated state, then to the metastable precipitation state and finally to the equilibrium state, than directly from solid solution to equilibrium state.

This can be shown as in Figure 42. Finally, a cascade reaction is more probable than a unique reaction.

CHAPTER 11

THEORY OF PRECIPITATION

11.1 General Considerations

As we have seen, observations made with the help of different metallographic techniques have established that hardening of alloys by ageing has as its origin the formation of fine precipitates. We propose in this chapter to give an account of the different theories of nucleation of these precipitates in the supersaturated solid solution.

Two types of nucleation can be distinguished: homogeneous nucleation and heterogeneous nucleation. Homogeneous nucleation, in theory, occurs only in a perfect crystal. The places where the precipitates are formed, are completely indeterminate and are distributed at random within the lattice. On the other hand, heterogeneous nucleation is produced at crystal defects. These defects are grain boundaries A, surfaces formed with dislocations B, isolated dislocations C and sometime inclusions D (oxides or carbides). These are places where precipitation is easy (Fig. 43).

A real crystal contains, between its defects, regions which are truly crystalline, the volume of which is sufficent for nucleation there to be homogeneous. We can then observe the two types of nucleation.

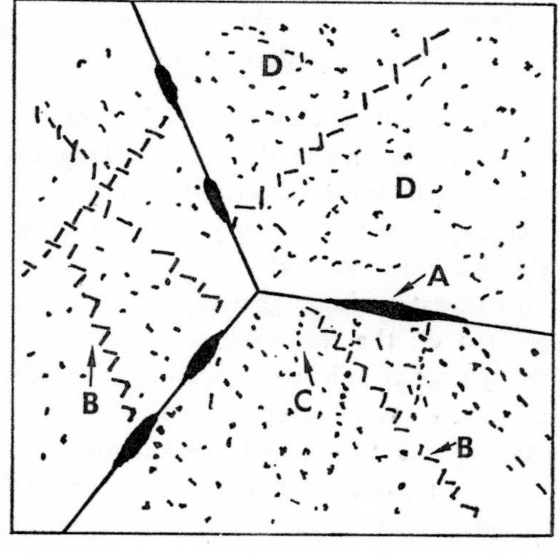

Fig. 43

11.2. Theory of Homogeneous Precipitation

Several theories have been suggested to explain the mechanism of homogeneous precipitation: Becker's theory of nucleation, Borelius' theory of fluctuations, and Hillert's theory of spinodal decomposition.

Before giving an account of the essential features of these theories, we will show that the solid solution has, before quenching, a structure which facilitates the formation of precipitated phases.

11.3. Structures of the Solid Solution in the Equilibrium State

An ideal solid solution is constituted of atoms distributed at random over the sites in the solvent. This leads on a very small scale to local variations of concentration in time and space, since homogeneity is a macroscopic property. It follows from this that an alloy brought to such a temperature that it is a homogeneous solid solution should be represented on the temperature-concentration graph preferably by a set of points situated inside an ellipse with its vertical axis corresponding to energy fluctuations and its horizontal axis to fluctuations in composition. In general the two types of fluctuation are connected.

If we consider the solution from the purely statistical point of view, the fluctuations in concentration result from the law of large numbers. The probability that a certain number of solute atoms are distributed over the sites inside a determinate volume may be calculated from Poisson's law. If the atoms are more numerous that given by the mean concentration, we have what is generally known as an embryo, because such groupings can serve as starting points for precipitation.

In a solid solution, embryos are continually being formed and dissolved, and do not reach a large size. The probability of their existence becomes smaller as their size increases. Figure 44 shows this result better than a calculation. The ordinate is the ratio R of the number of embryos

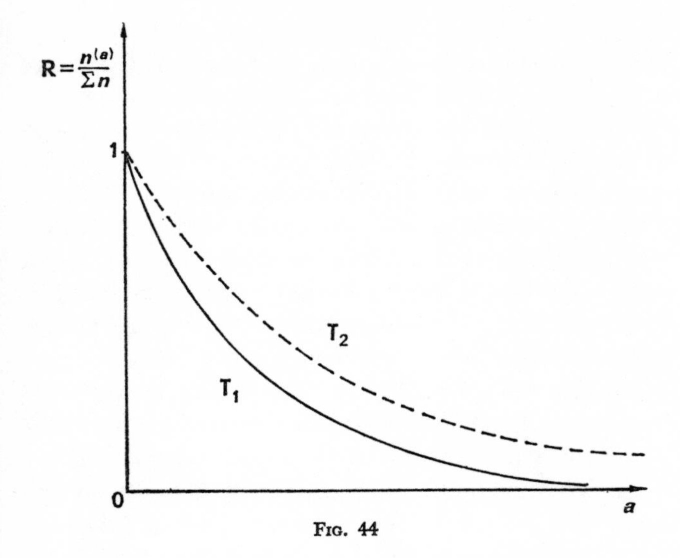

FIG. 44

containing a number a atoms of solute to the total number of embryos; the abscissa is the number of atoms a in one embryo. For a temperature T_1 the curve 1 corresponds to the pure statistical case of the ideal solid solution. The ratio R falls off very rapidly as a increases.

This also fits the case of a non-ideal solution if the temperature is much higher than the temperature limit of solubility for the concentration in question. The effect of temperature is then only felt in the rate at which the embryos are formed and redissolve.

If the solution is not ideal, and this becomes more

marked the nearer we are to the limit of solubility, the interactions between close neighbours depend on the nature of the atoms. We have seen that, from the first part of this book, if the interactions between similar atoms are stronger than those between dissimilar atoms, there is a tendency towards segregation, and that in the opposite case there is a tendency towards order. A local variation in concentration will always show itself by an increase in free energy, but this increase will be more or less marked according to the nature of the interactions, that is whether this change opposes or reinforces the tendency of the alloy.

We have drawn on Figure 44 a broken curve which corresponds to a non-ideal solution, held at a temperature T_2 slightly above the limit of solubility, having a tendency towards segregation. In this case the proportion of embryos containing a number of atoms a is larger than for T_1. If the alloy tended to move towards an orderly arrangement, on the other hand, the curve would be situated below the curve shown as a full line.

In addition to the effects of the interactions that we have just considered, it would be necessary to have regard to the effects of distortion of the lattice connected with the differences in size between solute atoms and solvent atoms. This might result in a change in form of the embryos, so as to minimise the free energy of the alloy. Calculation does show that the embryos are spherical in the absence of a distortion effect, but that they take the form of discs or needles when there is distortion.

Experiments have been carried out for some years with the object of demonstrating these local heterogeneities in the solid solution at high temperature. It appears that a positive result has been reached in some favourable cases (Al–Zn, for example).

To sum up, the quenched solution contains local variations in composition which can assist the formation of precipitates.

11.4. Force Tending Towards Precipitation

The formation of a precipitate in a supersaturated solution corresponds to a reduction in free energy, but this reduction is only the resultant of opposing influences. The equilibrium diagram shows that a supersaturated solid solution has a level of free energy greater than that of a mixture of the equilibrium phases. Also, the appearnce of a certain volume of precipitate or nucleus leads to a change in free chemical energy that we will call ΔF_1. This nucleus is in reality separated from the surrounding matrix by a geometrical or chemical interface to which there corresponds a surface energy ΔF_2. Precipitation may also lead to a change in atomic volume which will result in distortions of the lattices of the matrix and of the precipitate. It is therefore necessary to add a term for free energy of distortion, ΔF_3, linked with the precipitation.

The term ΔF_1 is negative, whilst the two terms ΔF_2 and ΔF_3 are positive since they correspond to the energy to be supplied to the alloy. For precipitation to take place, the following condition must be satisfied:

$$|\Delta F_1| > \Delta F_2 + \Delta F_3 .$$

We will see that there are several models of homogeneous nucleation. All of these involve the term ΔF_1 in the same way. The terms for surfaces and distortion, on the other hand, depend on the model chosen. Also, before giving an account of the three theories at present in competition, we will show how the term ΔF_1 can be evaluated when we know the curves of free energy as a function of concentration and temperature. Let us recall that these curves may be calculated or deduced from the equilibrium diagram.

11.5. Variation in Free Chemical Energy Connected with the Precipitation of a Phase

Let ΔF_1 be the change in free energy for a precipitate of n atoms, N the total number of atoms in the alloy, X the concentration of B atoms in the alloy, X_1 the concentration of the matrix in the neighbourhood of the precipitate, and X_2 the concentration in the precipitate.

We can put:

$$NX = (N - n)\, X_1 + nX_2$$

$$X - X_1 = \frac{n(X_2 - X)}{N - n}$$

$$\Delta F_1 = nf(X_2) + (N - n)\, f(X_1) - Nf(X),$$

where $f(X)$ is the free energy per atom of alloy at concentration X. In general, N is much greater than n and $X-X_1$ is very small. We can then write:

$$f(X_1) = f(X) - \frac{\mathrm{d}f}{\mathrm{d}X}(X - X_1),$$

whence:

$$\Delta F_1 = -\,n\,[f(X) - f(X_2) + (X_2 - X)\,f'(X)]$$

If we know $f(X)$ for all values of X, it is easy for us to calculate ΔF_1 for a precipitate containing n atoms.

The term in brackets may also be deduced graphically from the graph of free energy as a function of concentration, $f(X)$.

It is easily verified from Figure 45 that this term is equal to the length of the segment CD. Point D is on the tangent to the curve $f(X)$ for concentration X. We note with the help of this construction that ΔF_1 is at a maximum when the concentration corresponds to the point of inflexion of the curve $f(X)$. CD then has its highest value. Beyond this point the approximation that we have made is no longer valid.

Note. We have assumed that the concentration in the nucleus was equal to X_2, the concentration of the equilibrium phase. The result remains valid if we take, for example, a concentration intermediate between X and X_2.

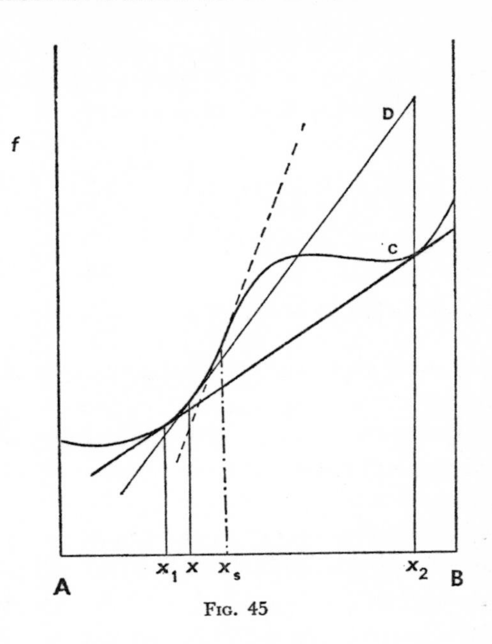

FIG. 45

11.6. Becker's Nucleation Theory

Becker had initially suggested this theory to explain the mechanism of condensation of vapours. He later adapted it to the case of alloys.

In this theory it is assumed that a certain number of nuclei are formed in the homogeneous supersaturated matrix, that these nuclei have the same composition but that their sizes are different. In addition, each nucleus can grow or shrink as a function of time. The change in free energy corresponding to the formation of each one of them is equal to the sum of the three terms already mentioned: ΔF_1, ΔF_2 and ΔF_3.

Since the first term is negative and the other two positive, formation of the nucleus may lead to an increase or a decrease in the free energy of the alloy. If however the increase in free energy is small, nucleation may be produced by an activation mechanism.

To show this, we consider the case of a supersaturated *regular solid solution* and we will suppose that the nuclei are cubic in form.

Let a be the side of such a nucleus and Δf the *change in free chemical energy per unit volume of precipitate* (this value is calculated starting from $f(X)$ and the number of atoms per unit volume). We will have:

$$\Delta F_1 = a^3 \Delta f.$$

If γ is the surface energy per unit area, then:

$$\Delta F_2 = 6a^2 \gamma.$$

Finally, if σ is the energy of distortion caused by unit volume of precipitate, we find:

$$\Delta F_3 = a^3 \sigma.$$

Hence for ΔF:

$$\Delta F = a^3 \Delta f + 6a^2 \gamma + a^3 \sigma.$$

ΔF is zero for $a=0$, and reaches a maximum value for:

$$\mathrm{d}\Delta F/\mathrm{d}a = 3a^2\Delta f + 12a\gamma + 3a^2\sigma = 0.$$

If a_0 is the corresponding value of a, we have:

$$a_0 = \frac{-4\gamma}{(\Delta f + \sigma)};$$

a cannot be negative.

When a is smaller than a_0, ΔF is positive and increases with a up to a value ΔF_0 for a_0, then falls again, becomes zero and then negative. This is shown on Figure 46.

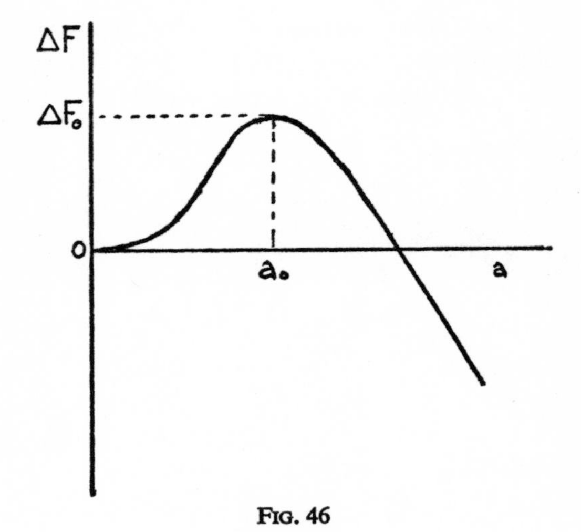

FIG. 46

Thus, below the critical size a_0, a nucleus will tend to disappear because an increase in size leads to an increase in the free energy of the alloy. Above this size, the free energy decreases with a.

Knowing a_0 it is easy to calculate ΔF_0:

$$\Delta F_0 = 2\gamma a_0^2 = \frac{32\gamma^3}{(\Delta f + \sigma)^2}.$$

The problem which arises then is to calculate γ and σ. It is difficult to calculate the energy of distortion, so let us take the case where it is negligible.

In order to evaluate Δf it is sufficient to consider the expression for the free energy of mixing of the regular solution:

$$\Delta G/N = X(1 - X)\,\omega + \\ + kT\,[X \log X + (1 - X)\log(1 - X)],$$

where ω is the energy of interaction which we calculated in Chapter VII.

To evaluate γ, we consider the bonding energies between close neighbours situated on opposite sides of the interface.

The reasoning is as follows:

Take two cubes of alloys, one with a concentration X_1, the other with concentration X_2. Then cut each of these cubes along a median plane and reconstitute two new mixed cubes as shown by Figure 47.

The change in surface energy ΔE_s, resulting from this

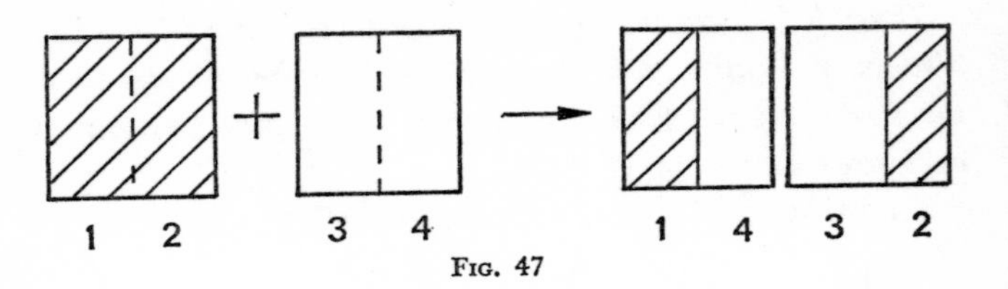

FIG. 47

operation, is given by the difference between the energy before and the energy after.

We can put $\Delta E_s = 2S\gamma$, if S is the area of the plane which is common to the two cubes.

We will assume that the atomic volume is the same for atoms of A as for atoms of B. The number of atoms per unit area is then equal to:

$$n_s = 1/V^{2/3}\,.$$

In the cube 1–2 the energy of surface S is equal to the sum of the bonding energies between the atoms situated on the two sides of S. The probability of having an atom of B on a site on one side of S is X_1. The probability that its neighbour on the other side of S is a B atom is also X_1, whence the number of BB bonds is:

$$(S/V^{2/3})\, X_1^2\,.$$

We find by the same reasoning that there are

$$S(1 - X_1)^2/V^{2/3} \quad AA \text{ bonds} \quad \text{and}$$

$$2SX_1(1 - X_1)/V^{2/3} \quad AB \text{ bonds}\,.$$

We carry out the same calculation for the other cube, then for the two mixed cubes. For the latter we have:

$$SX_1X_2/V^{2/3}$$
$$(S/V^{2/3})(1-X_1)(1-X_2)$$

and

$$(S/V^{2/3})[X_1(1-X_2)+X_2(1-X_1)]$$

bonds of type BB, AA and AB, respectively.

The increase in surface energy is therefore:

$$\begin{aligned}S\gamma = \frac{-S}{2V^{2/3}}\Big\{&4X_1X_2\frac{E_{BB}}{z}+4(1-X_1)(1-X_2)\frac{E_{AA}}{z}\\&+2[X_1(1-X_2)+X_2(1-X_1)]U_{AB}-\\&-2\frac{X_1^2E_{BB}}{z}-2\frac{(1-X_1)^2E_{AA}}{z}-\\&-2X_1(1-X_1)U_{AB}-2X_2^2\frac{E_{BB}}{z}-\\&-2(1-X_2)^2\frac{E_{AA}}{z}-2X_2(1-X_2)U_{AB}\Big\},\end{aligned}$$

where

$$U_{AB}=(E_{AA}+E_{BB}-\omega)/z;$$

$-U_{AB}$ is the bonding energy AB. This gives:

$$\gamma=2\omega(X_1-X_2)^2/zV^{2/3}.$$

In the case of the regular solution, we can determine ω if we know the critical temperature T_c, since, as we have seen:

$$\omega = 2kT_c.$$

The critical value for the side of a cubic precipitate then has the form:

$$a_0 = -\frac{8\omega(X_2 - X_1)^2}{V^{2/3}\Delta f} = -K\frac{(X_2 - X_1)^2}{\Delta f},$$

and for the change in free energy at the critical state ΔF_0:

$$\Delta F_0 = \alpha\frac{(X_2 - X_1)^6}{(\Delta f)^2}.$$

Thus Becker's theory enables us to determine the energy ΔF_0 which must be supplied to the alloy in order to form a precipitate of volume a_0^3.

ΔF_0 is the activation energy necessary to raise the free energy of the alloy. When the temperature is near to T_c, Δf is small and becomes zero when $T = T_c$. The activation energy of formation of the nucleus of critical size tends to infinity. In fact, one observes experimentally that in the neighbourhood of the solubility limit the rate of precipitation is very small.

The formation of a nucleus results in reality from the migration of the atoms. For nucleation to take place, the rate of diffusion must therefore be relatively large. This rate depends on the temperature and there is an activation energy of diffusion which corresponds to it. Let q be this activation energy per atom. The rate of nucleation is given by a law of Arrhenius:

$$I = I_0 \exp\left(-\frac{\Delta F_0 + q}{kT}\right),$$

q may be regarded as constant, whilst ΔF_0 depends on the temperature. As a result of this, I has a maximum for an intermediate temperature, because at low temperatures the diffusion is slow and this alone controls the reaction velocity. On the other hand, nucleation governs the kinetics of precipitation at high temperatures and then renders it very slow.

Figure 48 shows the trend of variation in activation energy of nucleation as a function of temperature, and gives two curves for I using two different activation energies of diffusion ($Q = Nq$). When the activation energy of diffusion increases, the maximum moves towards higher temperatures.

Note. In the preceding we have assumed, following Becker, that the concentration of the nucleus is constant and equal to the equilibrium phase which is precipitated. Since the publication of Becker's work, other authors have shown that the theory also applies if the concentration of the nucleus, whilst still remaining uniform, is not the same as that of the equilibrium phase. We can take any concentration between X_1 and X_2:

$$\Delta F_0 = \alpha (X_i - X_1)^6 / (\Delta f)^2$$

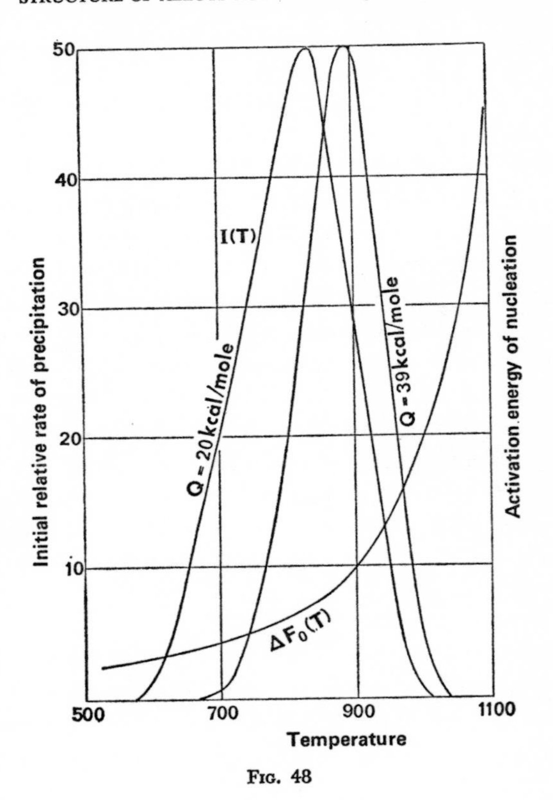

Fig. 48

will depend on the initial concentration and on the temperature. The nucleus most likely to be formed is that which requires the minimum activation energy.

11.7. Borelius' Theory of Fluctuations

Becker's theory regards the surface of the nuclei as of fundamental importance. It can be seen that such a theory can be easily applied to direct nucleation of a phase with a structure very different from that of the matrix within which it is formed. In many alloys, however, the first precipitates to appear have structures fairly close to those of the matrix. The role of the *common surface may therefore be less important* than in the other case. Borelius therefore considered that formation of the first precipitates could result from *fluctuations in concentration rather than from fluctuations in size*. As a simplification, he assumes that the nuclei *have all the same size, and that the concentration, although uniform, differs from one nucleus to another.* Despite this simplification he obtains some interesting results.

When one of these fluctuations takes place, the change in the free energy of the alloy results from the change in chemical energy, since the common surface and the distortions are not taken into account. To make the calculation from a knowledge of the curves of free energy as a function of concentration, it is sufficient to determine the change per atom of nucleus and to multiply by the number of atoms n. The reasoning is the same as before, but in place of treating the concentration in the precipitate as constant, it is treated as a variable. If X_g is the concentration of atoms in the domain, and X_i the initial concentra-

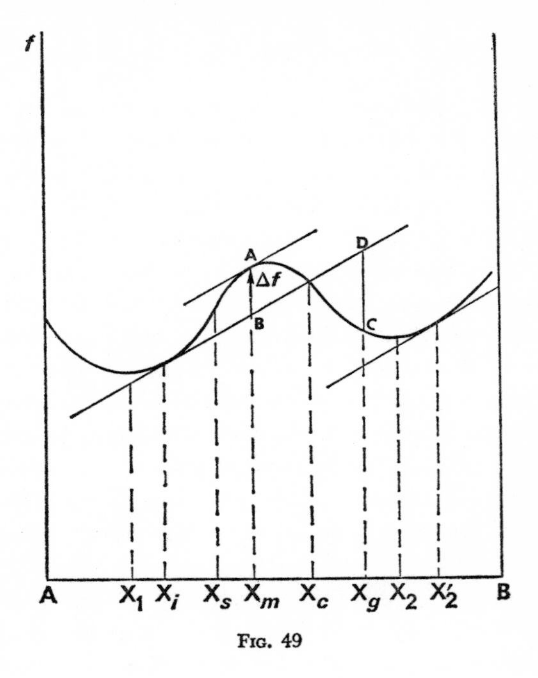

FIG. 49

tion of the alloy, we can put:

$$\Delta F = n\,[f(X_g) - f(X_i) - (X_g - X_i)\,\partial f/\partial X_i]\,.$$

To see how ΔF changes with X_g, consider Figure 49.

We have seen that, for a concentration X_g, the change in chemical free energy per atom is given by the segment DC on this graph. It is clearly seen that when X_g changes between X_i and X_2 ,the magnitude of DC changes as does its sign. The change in chemical free energy is negative beyond X_c, the point of intersection of the tangent at X_i with the curve. It is positive when X_g lies between X_i and X_c. Hence nucleus formation is accompanied by a reduction in free energy when $X_g > X_c$.

Borelius' mechanism corresponds to formation of domains where the concentration increases regularly. We see that if a domain reaches concentration X_c, any further growth leads to a reduction in the free energy of the alloy.

Before concentration X_c is reached, the free energy increases with X_g, but the increase reaches a maximum Δ_{fm}, given by point X_m, where the tangent to the curve $f(X)$ is parallel to the tangent at X_i. It is also evident that the maximum value of Δf_m depends on the value of X_i. This is greatest when $X_i = X_1$, and on the other hand it is zero when $X_i = X_s$, where X_s is the value of the concentration at the point of inflexion (spinodal concentration).

Figure 50 shows the change in Δf as a function of X_g for three values of the initial concentration. One of these curves corresponds to $X_i < X_s$, another to X_i between X_i and X_s, and the third to X_s. In the alloy with initial concentration X_i the free energy increases so long as the composition does not reach the value X_m in its fluctuations, and it decreases as soon as X_m is exceeded. Therefore, according to this theory there is an activation barrier to be passed for precipitation to take place.

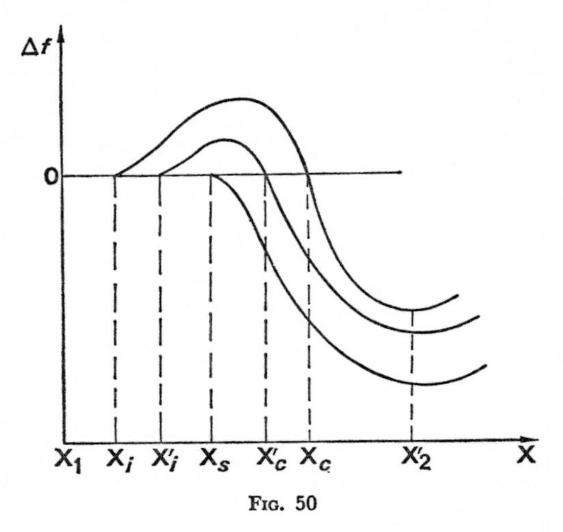

FIG. 50

This leads to an activation energy of precipitation:

$$W = n\Delta f_m ,$$

where n is the number of atoms in the nucleus.

For an alloy of initial composition X_s, $\Delta f_m = 0$ and there is no activation energy.

The probability of appearance of a nucleus being proportional to the probability of fluctuation of concentration X_m and to the diffusion coefficient, we find for the initial precipitation rate a formula analogous to Becker's:

$$I = I_0 \exp\left[-\left(Q + n\mathcal{N}\Delta f\right)/RT\right].$$

The initial rate would be infinite inside the spinodal

domain. A small fluctuation in concentration tends to disappear if X_i is smaller than X_s, and it becomes more pronounced if $X_i > X_s$.

Note the simplicity of this model. It is assumed that the concentration is uniform in the matrix and in the domain where the fluctuation is produced.

11.8. Comparison Between Becker's and Borelius' Theories

We have seen the special part played, in the two theories, by the spinodal concentration. In Becker's theory this concentration corresponds to the maximum value of the activation energy of nucleation and in Borelius' theory to a zero value of this energy.

The two reasonings are only comparable within the domain outside X_sX_s'. Inside the X_sX_s' region fluctuations are formed immediately whilst nucleation must be activated.

Hobstetter and Scheil have attempted to make a synthesis of the two theories in the case where $X_1 < X_i < X_s$. They are of opinion that neither model must be discarded, but that nucleation takes place according to the easiest process, that which requires the smallest activation energy.

If we graph the change in energy of nucleation as a function of the size of the nucleus for the two theories, we obtain the result given by Figure 51.

In the latter case, ΔF is obtained by altering the number

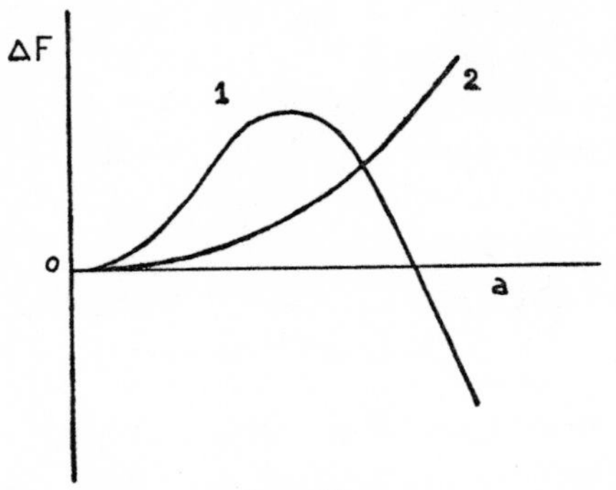

FIG. 51a

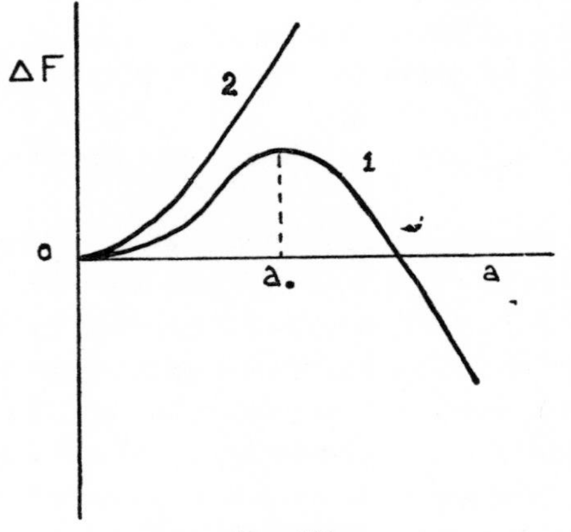

FIG. 51b

of atoms in the domain, for a value Δf_m given by the initial concentration of the alloy.

In both cases the form of the curve ΔF(a) depends on X_i. From this there result two types of configuration as shown by Figures 51a, b.

In the case of Figure 51a, formation of a nucleus is more probable by fluctuation than by creation of an interface. When, however, the size of this nucleus exceeds that corresponding to the intersection of the two curves, growth with an interface is favoured. A fluctuation is transformed into a precipitate separated from the matrix by an interface.

For Figure 51b, formation of a nucleus surrounded by an interface is more easy that formation of a fluctuation of the same size.

When X_i reaches X_s, fluctuation becomes of course the easiest mode of decomposition.

11.9. Spinodal Decomposition

We have seen the importance given to concentration X_s in the theory of fluctuations. An alloy with a concentration lying between X_s and X_s' (spinodal concentrations) decomposes without activation energy of nucleation. Any fluctuation in concentration then tends to grow and we will see that the quenched structure is transformed during ageing into a modulated periodic structure.

Let us suppose as a simplification that the decomposition of the matrix is not accompanied by distortions of the lattice and that it leads to formation of two terminal solid solutions of the same structure and of compositions X_1 and X_2.

The curve of free energy as a function of concentration is of the same type as that of Figure 45.

The concentration of the quenched alloy is not entirely uniform due to statistical fluctuations and to the divergence from the ideal state. *Now a local variation in concentration is equivalent to a diffuse interface.*

To show this, let us suppose that the variation of concentration is by planes. Under these conditions an atomic plane will have a uniform concentration, but this concentration will vary from one plane to another. Each plane may be regarded as a nucleus of uniform concentration X, bounded by two planes of concentrations $X+\Delta X$ and $X-\Delta X$.

According to Becker the equivalent surface energy is proportional to $(\Delta X)^2$. The local free energy will include the term $f(X)$ and a term proportional to $(\Delta X)^2$.

In the general case of a change in concentration of any form whatsoever, the local free energy will be the sum of the term $f(X)$ and a term proportional to the square of the concentration gradient ∇X.

To obtain the free energy F of the crystal, it is necessary to find the sum of the local free energies.

If n_v is the number of atoms per unit volume, we have:

$$F = \int_V n_v [f(X) + K(\nabla X)^2] \, \mathrm{d}V .$$

This expression is in reality only an approximation and derivatives of higher order would need to be brought in. The problem now set consists of comparison of the free energy of the crystal, within which the concentration is not uniform, with that of a crystal where the concentration is taken as uniform. Let X_0 be the mean concentration of the alloy and X the local concentration. X depends on the co-ordinates (u, v, w) of the point of the crystal under consideration.

The difference in free energy between the two configurations is given by:

$$\Delta F = n_v \int_V [f(X) + K(\nabla X)^2] \, \mathrm{d}V - n_v \int_V f(X_0) \, \mathrm{d}V$$

$$\int_V f(X_0) \, \mathrm{d}V = V f(X_0).$$

Since the local variations $X - X_0$ are small in size, we can develop $f(X)$ as a series:

$$f(X) = f(X_0) + (X - X_0) \frac{\partial f}{\partial X_0} + \tfrac{1}{2}(X - X_0)^2 \times \\ \times \frac{\partial^2 f}{\partial X_0^2} + \cdots,$$

which gives:

$$\Delta F = n_v \int \left[(X - X_0) \frac{\partial f}{\partial X_0} + \tfrac{1}{2}(X - X_0) \frac{\partial^2 f}{\partial X_0^2} + \\ + K(\nabla X)^2 + \cdots \right] \mathrm{d}V;$$

$\partial f / \partial X_0$ and $\partial^2 f / \partial X_0^2$ do not depend on u, v, and w.

We can describe the concentration profile of the crystal as a sum of sinusoidal functions and we can calculate the contribution to the change in free energy ΔF made by each of them.

Suppose the change of concentration to be unidirectional, we have:

$$X - X_0 = A \cos(\beta\mu) .$$

This gives in turn:

$$\int_V (X - X_0) \, \mathrm{d}V = 0$$

$$\int_V (X - X_0)^2 \, \mathrm{d}V = \tfrac{1}{2} A^2 V$$

$$\int_V (\nabla X)^2 \, \mathrm{d}V = A^2 \beta^2 \int_V \sin^2(\beta\mu) \, \mathrm{d}V = \tfrac{1}{2} A^2 \beta^2 V ,$$

whence:

$$\Delta F = n_v V \frac{A^2}{4} \left(\frac{\partial^2 f}{\partial X_0^2} + 2K\beta^2 \right) .$$

The stability of the uniform solution with regard to the modulated periodic structure depends on the sign of ΔF.

Note that K is always positive, since $K(\nabla X)^2$ is the surface energy.

Let us consider the different positions for X_0:

(1) $X_0 < X_s$ or $X_0 > X'_s$:

$$\partial^2 f / \partial X_0^2 > 0 \quad \text{therefore} \quad \Delta F > 0 .$$

The uniform solid solution is more stable than any sinusoidal fluctuation.

(2) $X_s < X_0 < X_s'$:

$$\partial^2 f/\partial X_0^2 < 0 .$$

If the wavelength of the modulation $2\pi/\beta$ is shorter than $2\pi/\beta_c$, given by $\partial^2 f/\partial X_0^2) + 2K\beta^2 = 0$, $\Delta F > 0$, the uniform solution is stable.

If $\beta < \beta_c$, $\Delta F < 0$, the corresponding periodic modulation is more stable than the uniform concentration.

When the crystal is of infinite dimensions, β varies from 0 to infinity by discrete values, therefore β_c will be one of the range of wavelengths present in the crystal. The uniform solution will be unstable as soon as X_0 becomes equal to X_s.

If the crystal is of limited size, β will have a non-zero lower limit $2\pi/L$ and instability will start as soon as X_0 is such that:

$$\partial^2 f/\partial X_0^2 = - 8\pi^2 K/L^2 .$$

Thus, the formation of a periodic modulated structure in a small crystal depends on the relative value of the concentration X_0 and can only take place inside the spinoidal curve, the position of the points X_s and X_s' as functions of temperature.

From what has been said, we see that the contribution of a harmonic component to ΔF will be more negative the smaller the corresponding term β. In consequence, the evolution of the structure with time will proceed in such a manner that long-wave harmonics will become preponderant. A coalescence phenomenon is produced. This could also be predicted by calculation when we solve the differential equations for diffusion by inserting the chemical potential gradients. In this case we find that the coefficient of diffusion is proportional to $\partial^2 f/\partial X_0^2$, therefore the direction of the concentration change depends on the sign of this term; we also find that the wavelength spectrum inside the spinodal domain comprises harmonics which become larger as the ageing time increases.

In the preceding, we have assumed that the solid solution was an isotropic medium and that the change in composition did not bring with it any elastic deformation of the lattice. In actual fact, in most alloys, the lattice has anisotropic elastic properties and decomposition is accompanied by elastic deformations of coherence.

We can make allowance for these supplementary effects by inclusion in the expression for F of a term linked with the anisotropic elastic deformation due to local variation of concentration $X-X_0$:

$$F = n_v \int_V [f(X) + H(X - X_0)^2 + K(\nabla X)^2]\,\mathrm{d}V,$$

which gives for ΔF:

$$\Delta F = \tfrac{1}{4}A^2\left(\frac{\partial^2 f}{\partial X_0^2} + 2H + 2K\beta^2\right) n_v.$$

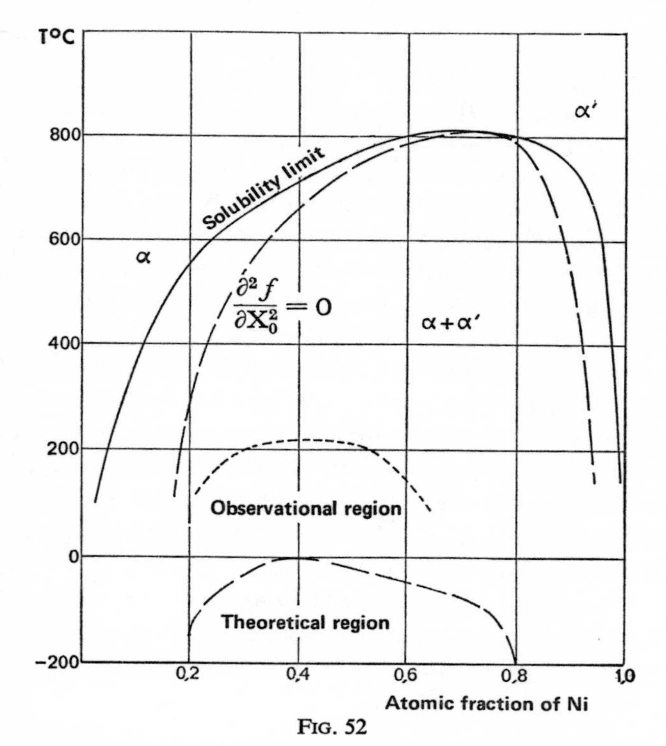

FIG. 52

The result is a displacement of the limit of instability towards the interior of the spinodal domain and an isotropy in the concentration profile.

For Au–Ni alloys, for example, instability can exist only at very low temperatures. One finds a spinodal curve, that of elasticity, shown on Figure 52.

We see on this graph that the chemical spinodal curve is close to the limiting solubility curve and very different from the spinodal elasticity curve. The alloy cannot decompose spontaneously at a temperature above the surrounding temperature and direct formation by heterogeneous nucleation of the equilibrium phase is facilitated. In fact, electron diffraction has shown that homogeneous decomposition is produced from 200° C, which proves that the theory of spinodal decomposition does not yet take into account all the factors concerned.

The theory of spinodal decomposition is therefore able to explain the formation of the periodic structures of which we have spoken above. By taking account of the elastic anisotropy of the crystal, it predicts the formation of *preferred* directions of modulation which correspond to observation of discs or of cubes parallel to certain crystallographic planes in the lattice.

Despite the simplifying assumptions made, this theory provides very important new features for the understanding of the phenomena of pre-precipitation.

CHAPTER 12

GROWTH OF PRECIPITATES FORMED BY HOMOGENEOUS NUCLEATION

When the nuclei have exceeded a certain critical size, whatever value this may have, they may be regarded as precipitates which will develop with the help of solute atoms removed from the supersaturated matrix.

For homogeneous precipitation, the precipitates are distributed at random and we can consider three situations: the first corresponds to precipitates which are small in comparison with the distance separating them from their close neighbours. Growth will take place for a certain time as if they were alone in the crystal. The second situation will occur when the precipitates have removed enough B atoms from the lattice for this no longer to have the initial composition c_0 at any place between the precipitates. Finally, the third situation corresponds to the evolution of the precipitates after the matrix has reached its minimum composition in B atoms at all points between the precipitates. In this case we come to the process of coalescence, leading to growth of the larger precipitates at the expense of the smaller.

Figure 53 represents schematically the three situations. It shows the concentration profile through the precipitates, which can equally well be spheres, cubes or discs.

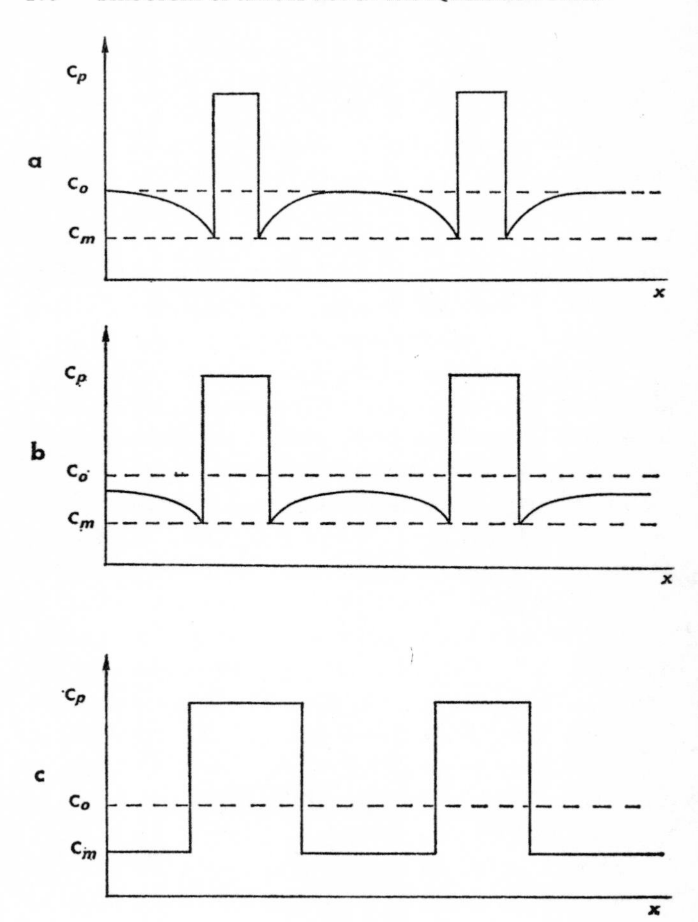

FIG. 53

Each of these processes has its own particular kinetics. We will not give the corresponding calculations, but will only state the conclusions so as to show the laws arrived at, and we will compare them with some experimental results.

12.1. Growth of Precipitates Isolated in the Matrix

We must still consider two extreme cases. Growth can be controlled either by diffusion of atoms in the lattice or by passing over the interface. For a nucleus to develop, the solute atoms must leave their initial positions in the lattice and move towards the precipitate, but they must also pass over the interface.

The influence of the reaction at the interface is appreciable only at the start of growth. When the precipitate is small, growth is produced by the closest atoms, which have a very short diffusion time. On the other hand, when the precipitate is large, atoms must travel a long distance before being incorporated into the precipitate. Growth is then controlled by the phenomenon of diffusion.

Case of reaction at the interface

Since diffusion is rapid, we can assume that the concentration of element B in the matrix remains substantially uniform. Let C_f be the concentration in the matrix which has lost atoms and C_g the concentration in the nucleus (this corresponds practically to equilibrium).[1] We can assume that the gain g_B in atoms is proportional to the difference in concentrations at the interface, at time t and at the end when C_f is reached:

$$g_B = \alpha(C_t - C_f),$$

where C_t is the concentration in the matrix at time t.

The increase in volume $\mathrm{d}V$ of the precipitate corresponds to a gain of $\mathrm{d}V(C_g - C_f)$ atoms, which will therefore be equal to $g_B\mathrm{d}t$, whence:

$$\frac{\mathrm{d}V}{\mathrm{d}t} = \alpha\frac{(C_t - C_f)}{(C_g - C_f)}.$$

We can assume that the concentration in the matrix decreases and we express it as a mean by the fraction of atoms precipitated, y. We then have:

$$1 - y = \frac{(C_t - C_f)}{(C_0 - C_f)}.$$

where C_0 is the initial concentration in the matrix, whence:

$$\frac{\mathrm{d}V}{\mathrm{d}t} = \frac{\alpha}{C_g - C_f}[(C_0 - C_f)(1 - y)].$$

The result of this is that the rate of growth of the precipitate decreases continually with time.

To solve the problem we need to know $y(t)$. We can assume that y is negligible during the first moments of the reaction, corresponding to a constant rate according to the above formula.

The reaction at the interface is thermally activated and α has the form of an Arrhenius law. This growth mechanism is difficult to observe, since it occurs when the precipitates are extremely small and widely dispersed.

Case where diffusion controls growth

The reaction at the interface is then much more rapid. We assume that the concentration in the lattice at the interface is equal to C_f and that the concentration profile in the lattice is as shown in Figure 53a.

Suppose the interface to be plane. When the size increases, it moves a distance $\mathrm{d}R$ and there are $\mathrm{d}R(C_g - C_f)$ atoms of B entering the precipitate per unit area. This flux is equal to that which supplies the interface from the reservoir formed by the lattice. It can be expressed by Fick's law, and if r is the distance within the lattice at which we determine C, we obtain:

$$(C_g - C_f)\frac{\mathrm{d}R}{\mathrm{d}t} = D\left(\frac{\partial C}{\partial r}\right)_R .$$

$(\partial C/\partial r)_R$ corresponds to the concentration gradient of the matrix at the interface. To solve the equation we must assume a law for the increase of R with time, then verify whether the solution fits. We can make the approximation that R is small compared with the maximum distance to be traversed by the atoms, that is the zone where the lattice is impoverished. This means that $\mathrm{d}R/\mathrm{d}t$ can be neglected and we can take $R = R_1$. We then have a stationary state. The change in concentration is given by the solution of the following equation which we will not prove:

$$\frac{\partial C}{\partial r} = \frac{C_{R_2} - C_f}{1/R_1 - 1/R_2} \frac{1}{r^2}.$$

Taking the values at the limits: $r = R_1$ and $r = R_2$.

C_{R_2} is the concentration for $r = R_2$ in the case of the spherical precipitate, R_1 is the radius of the precipitate.

Hence we obtain, if $R_2 \to \infty$ and $C_{R_2} = C_t$, the concentration in the matrix:

$$\frac{R_1 \mathrm{d}R_1}{\mathrm{d}t} = D\left(\frac{(C_t - C_f)}{(C_g - C_t)}\right).$$

There are, in fact, many nuclei and the zones where the matrix is impoverished overlap. It is therefore necessary to have regard to these overlappings and to apply a correction proportional to the non-transformed fraction. We can put:

$$\frac{C_t - C_f}{C_0 - C_f} = 1 - y,$$

since C_t does not greatly differ from the mean concentration at time t. Hence:

$$R\frac{\mathrm{d}R}{\mathrm{d}t} = D\left(\frac{C_0 - C_f}{C_g - C_f}\right)(1 - y).$$

To integrate this differential equation, we assume that

at the start of the reaction y is not greatly different from 0. We then obtain:

$$R^2 = kt \quad \text{and} \quad V = k' t^{3/2},$$

where V is the volume of precipitate.

This theory is valid only for a small supersaturation for which the dimensions of the impoverished zone within the matrix are much larger than the precipitate. If the degree of supersaturation is large, it is no longer possible to assimilate the diffusion phenomenon to a stationary state.

In order to give a rigorous treatment of this subject, it would be appropriate to have regard to the anisotropy of shape of the precipitate, to diffusion and stress in the lattice, but this greatly complicates the theory.

To determine the nature of the reaction kinetics for precipitation in an alloy, regard must be paid to the number of precipitates, in addition to their individual growth. This leads to the introduction of the rate of nucleation and by so doing complicates the problem still further. As a simplification we consider extreme cases. For example, we suppose that nucleation is so rapid that all the nuclei are formed simultaneously. It is then sufficient to multiply the number of nuclei by the rate of growth of one precipitate. We can also assume a law of nucleation and integrate with respect to time the individual growths of all the nuclei.

We see that these theories are based on many simplifying assumptions. To make progress it is appropriate to design more sophisticated models of growth. The theories have, in fact, always been indirectly compared with experiment and agreement has rarely been satisfactory.

12.2. Coalescence of Precipitates

When the precipitates are very numerous, they cannot grow indefinitely because their growth takes place at the expense of the matrix, which is of limited volume and which rapidly becomes impoverished. Due to the nucleation which takes place with time, certain precipitates are larger than their neighbours. There is a fairly wide distribution of sizes, but there is however a distinct maximum. The dispersed state does not correspond to the minimum free energy of the system. In fact the surface energy becomes larger as the precipitate, at equal volume, becomes smaller, and equilibrium is only stable when only one precipitate co-exists with the impoverished matrix.

It is easy to see that the surface of a sphere of volume V is less than that of two spheres each of volume $V/2$. There will therefore be a tendency for small-sized precipitates to disappear and large-sized precipitates to be formed, at approximately equal volume of precipitated material. This is what is known as the phenomenon of coalescence. It takes place earlier as the alloy becomes more strongly supersaturated, and more rapidly as diffusion becomes easier, therefore at higher ageing temperature.

We will not be able to develop the theory of coalescence, which without being very complicated is lengthy. We will limit ourselves to showing that it rests on the change of concentration of the matrix in the neighbourhood of a precipitate when the concentration of the latter alters; we will then give without proof the formulae reached for distribution of sizes and for the law of average growth.

Let us consider a spherical precipitate of radius r inside a matrix. If this precipitate is not to dissolve spontaneously the matrix must have a certain composition. The chemical potential of each of the components must be the same in the matrix and in the precipitate. This chemical potential is equal to $\partial G/\partial N_B$ for element B. Since G, for this precipitate, contains a chemical term as well as a surface term (we still suppose that there is no energy of deformation due to the precipitation), the chemical potential will be different from that of the thermodynamic equilibrium of two phases for which surface energy is not a factor.

We can calculate the relationship connecting r with the activity of the component B in the solid solution without knowing the expression for G. Let a_0 be the activity corresponding to equilibrium, that is when r tends to infinity, and a_r the activity for a radius r, we can put the change in free energy $\mathrm{d}G$ due to transference of $\mathrm{d}N$ atoms of B from the particle to the equilibrium lattice supposed to exist far inside the crystal:

$$\mathrm{d}G = \mathrm{d}NRT \log(a_r/a_0),$$

since the atoms pass from the phase where the activity is a_r to that where it is a_0.

The change in corresponding area of the particle is:

$$dS = 8\pi r \, dr,$$

and putting v as the volume per mole:

$$v \, dN = 4\pi r^2 \, dr,$$

whence:

$$dS = 2(v/r) \, dN.$$

The change in surface energy is equal to γdS, if γ is the energy per unit area of the precipitate surface. Since the surface energy contributes to the increase in free energy, we can put:

$$\gamma dS = dNRT \log(a_r/a_0),$$

whence:

$$\boxed{RT \log(a_r/a_0) = 2v\gamma/r.}$$

In the neighbourhood of a particle of radius r, the activity must be changed from a_0 to a_r so as to prevent the particle from being dissolved for the benefit of a very large particle near which the activity does not greatly differ from a_0.

Let us now suppose that we have two particles of different radii placed close together but not in contact. Since

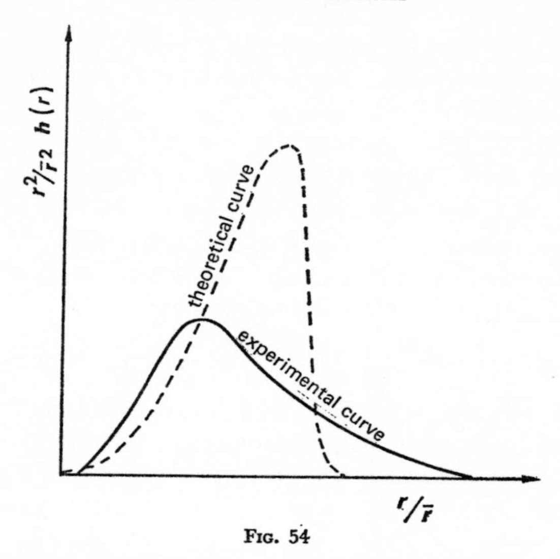

Fig. 54

they are of different sizes, the matrix does not have the same activity in their respective neighbourhoods. There is therefore a concentration gradient, leading to diffusion of atoms from the neighbourhood of the smaller towards the larger. As a result there is a reduction in the concentration of the matrix around the small particle and hence it dissolves, whilst the large particle increases in volume.

Starting from these considerations, we can calculate the evolution of a system containing particles where the size obeys a given distribution law. We find that the distribution of sizes rapidly becomes stationary and that it has a theoretical form as given in Figure 54.

The ordinate is the frequency $h(r)$ multiplied by the ratio $r^2/\bar{r}^2$, where $\bar{r}$ is the mean value of r, and the abscissa is the ratio $r/\bar{r}$.

This method of representation has been chosen for theoretical reasons that we will not give in detail.

We also find that if r_m is the mean diameter of the particles, those of diameter $2r_m$ grow more rapidly than all others, whilst those of diameter less than r_m are dissolved.

Calculations lead to a law of growth of form:

$$r_m^3 - r_0^3 = \alpha t\,,$$

where α is a constant depending on the temperature and on the equilibrium concentration in the matrix.

Numerous experiments have been carried out during the last ten years with the object of verifying this theory in the cases where it was certain that coalescence took place, and in other cases in order to show the existence of coalescence. Figure 54 shows that the observed distribution function for diameters differs considerably from the theoretical It appears, however, that its form remains stationary as predicted by the calculation. Better agreement has been obtained for the law of growth of mean diameter as a function of time, $r_m^3 = r_0^3 + \alpha t$.

If we plot the cube of the mean radius against the time, this mean radius being measured by means of the electron microscope, we obtain a straight line as predicted by theory, with its gradient depending on the temperature. If we calculate the corresponding heat of activation, we

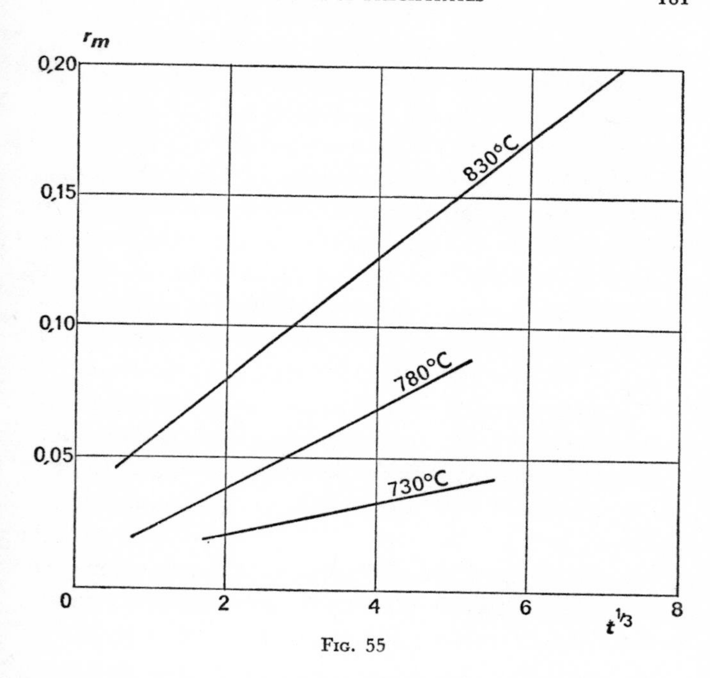

FIG. 55

find in general the heat of activation for diffusion in the matrix.

Figure 55 shows as an example the curves obtained for iron–copper alloys containing 2% and 6% copper.

The law is obeyed for the three temperatures. Further, for a given temperature the points corresponding to each alloy lie on one single straight line.

In the preceding we have assumed that the phenomenon

of coalescence was controlled by the rate of diffusion in the lattice. This implies that the reaction at the interfaces is extremely rapid. In the opposite case, diffusion is very rapid and the growth law obtained is different. It does not appear that this situation can be met with in precipitation in alloys. In any case, it could only happen at the very beginning of precipitation, when the particles are very small and tightly packed, so that the range of diffusion is very much limited. It is then very difficult or even impossible to verify the law of coalescence by practical means.

Lastly, we have assumed that the particles dissolving or growing were spherical. In practice, as we have seen already, they may be cubic or more frequently disc-shaped. We can nevertheless modify the equations to give results not very different from those found for the simple case of spheres.

Bearing in mind the difficulty of verification for this latter type, it does not appear urgent to seek a more detailed theory with the aim of complete generality.

NOTE

[1] The concentrations correspond to the number of *B* atoms per unit volume.

CHAPTER 13

DISCONTINUOUS PRECIPITATION

13.1. General Considerations

As we have previously stated, heterogeneous precipitation differs from homogeneous precipitation in the distribution of the nucleation sites. In the case of heterogeneous precipitation, nuclei are formed at faults in the structure, of which the activity from this point of view may be very different, depending on their nature or on the conditions of treatment, temperature and time of ageing.

The precipitation reaction is then produced by growth starting from each nucleus of the transformed volume. The general term given to this type of reaction is 'discontinuous precipitation'. The adjective 'discontinuous' is used because the composition of the matrix changes abruptly when it is affected by the reaction. Progress of the reaction takes place by displacement of a front which separates the transformed volume from the volume not yet transformed.

In the transformed zone, the precipitated phase and the impoverished matrix exist together. The composition of this latter is uniform and corresponds effectively to thermodynamic equilibrium. As a result, crossing the boundary corresponds to a discontinuous change in the properties of the matrix. In particular when we measure the size of a unit cell of the lattice of a sample in course of transformation, we obtain two distinct values: one is characteristic of the unchanged lattice and the other of the impoverished lattice.

Discontinuous precipitation can be produced by means of two processes: discontinuous precipitation with recrystallisation and discontinuous precipitation without migration of grain boundaries.

13.2. Discontinuous Precipitation with Recrystallisation Reaction

This reaction may be considered as of perlitic type. In fact, it is generally produced in alloys with eutectoid transformation, of which the commonest and most important is the iron-carbon system. When an iron alloy containing about 0.8% by weight of carbon is brought to 900°C, a solid solution of carbon is formed in the face-centred cubic iron. If it is cooled relatively slowly from this temperature, polished and etched, a complex structure is observed under the optical microscope. This consists of grains of ferrite deeply embedded in each other, and fine laminae of cementite (Fe_3C) which give the whole system a pearly appearance. It is this appearance, resembling mother-of-pearl, which leads to the name of perlite.

Figure 56 shows schematically the appearance of perlite

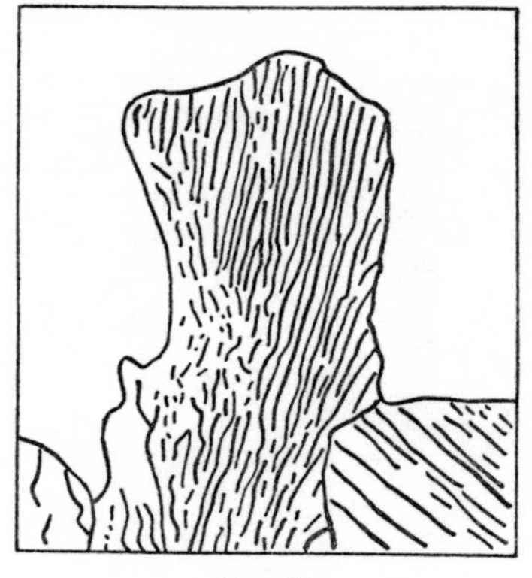

FIG. 56

in a partially transformed zone. The orientation of the plates is well defined with regard to the lattice of the grain containing them, and varies from one grain to another.

The plates of cementite are finer and more closely packed the greater the rate of cooling. Beyond a certain rate, they can no longer be separated and their existence is revealed only by the change in colour from one grain to another. This type of structure is also obtained after quenching and ageing.

The characteristic feature of this process of precipitation is its growth from the grain boundaries. After a few moments ageing, the grain boundaires are covered with ribbons which develop alternately in one grain or in the other, as shown schematically in Figure 57. These ribbons are due to local displacement of the boundary which is de-

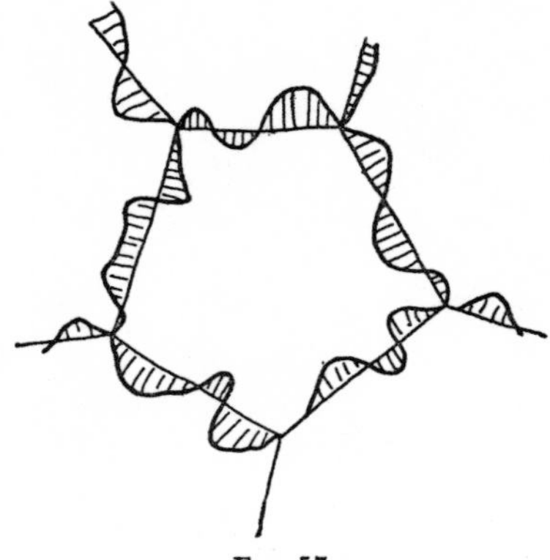

Fig. 57

formed. The transformation is complete when the whole of each grain is covered by these ribbons.

This type of precipitation only occurs at very large degrees of supersaturation. This was the case with the iron–carbon eutectoid, and is also the case with alloys forming, at high temperatures, extended solid solutions such as Au–Ni, Ni–Ti, and Fe–Zn. In certain of these alloys, discontinuous precipitation is very rapid at low temperatures and slow at high temperatures; the rate of diffusion in the grains is then sufficient to allow homogeneous precipitation to take place in advance of the reaction front; this tends to hinder the discontinuous reaction. This is because the change in free energy, which is the driving force, becomes smaller.

The two-phase region of the equilibrium diagram may

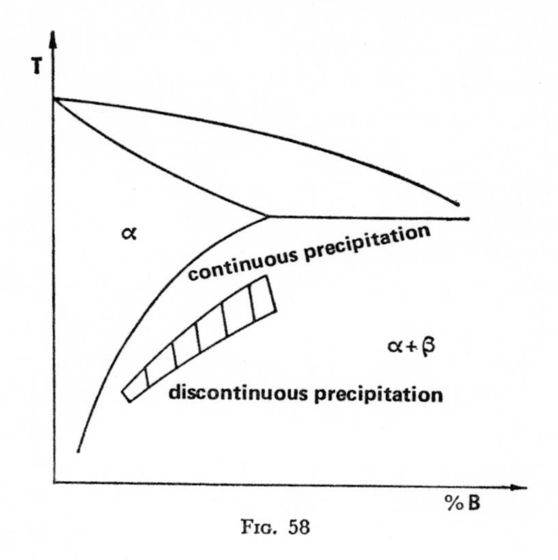

FIG. 58

thus be separated into two parts: one where only homogeneous precipitation is observed, the other where only the discontinuous reaction takes place. Figure 58 shows schematically this effect. There also is a region where both types are observed, either in succession or simultaneoulsy. The discontinuous reaction, then, is never complete.

13.3. Discontinuous Precipitation without Displacement of Grain Boundaries

This reaction has been observed to a fairly small extent. It takes place by direct nucleation of the equilibrium phase which is precipitated on the isolated dislocations and on the dislocations which form the subgrain boundaries (disorientation less than 10°). It progresses by development of dendrites in the solid matrix, but this takes place along a front which is relatively plane.

The dendrites have a well-defined orientation with respect to the matrix lattice. Thus, in an alloy of iron and molybdenum containing 25% of molybdenum by weight, dendrites are observed in the $\langle 110 \rangle$ directions. The growth front is a $\{101\}$ plane. The relationships between the orientations of the precipitated hexagonal phase and the lattice behind the front are complex but well-defined.

Along the front where the dendrites progress, the solute atoms are drained away through a limited volume where the concentration is not uniform. Behind this in the matrix, the concentration corresponds almost to that of equilibrium. Nucleation at the dislocations is extremely rapid. The rate of displacement of the front is governed by the volume diffusion. Its speed is constant.

The reaction rate depends on the distribution of the dislocations: isolated dislocation lines, surfaces of subgrain boundaries. These latter are, in fact, composed of aligned dislocations at short distances apart, so that the zones issuing from each dislocation touch each other almost immediately and so form only two fronts of progression one on each side of the plane of the sub-grain boundary.

This type of reaction has been observed outside the

Fe–Mo system in alloys of Fe–W, Fe–Be and in FeO, which may be regarded as an alloy containing iron vacancies in solution.

Lastly, we note that a discontinuous reaction can be followed by another discontinuous reaction. If the reaction takes place at low temperatures, the precipitates or the dendrites are very fine. The state is still metastable. There can be either normal coalescence or a reaction of perlitic type leading to a coarser structure, with a low activation energy. Such a sequence of discontinuous reactions has been observed in Fe–Ni–Ti alloys, aged in the region of 550° C.

13.4. Mechanism of Discontinuous Precipitation

Diffusion of the atoms in an alloy can be produced in two ways: intergranular diffusion and intragranular diffusion.

Intragranular diffusion necessitates a relatively high activation energy, and takes place at high temperatures. On the other hand, intergranular diffusion takes place at low temperatures since the atoms move more readily along the grain boundaries than in the crystals. We thus see that discontinuous precipitation is favoured at low temperatures. As they move, the boundaries sweep through the matrix and facilitate the passage of solute atoms in the precipitates.

Although the energy of the boundary is thus increased, the gain in chemical free energy due to precipitation constitutes a driving force which is sufficient for the reaction. Moreover, the activation energy of the process is smaller than that corresponding to homogeneous precipitation.

13.5. Kinetics of the Reaction

A large amount of study and research has been devoted to discontinuous precipitation. It does not appear, however, that a complete solution to the problem has been reached. Most of the theories are phenomenological and we will summarise here the essential features of the more important ones.

As with homogeneous precipitation, two stages of the reaction must be considered: nucleation, then growth.

As we have said, nucleation takes place at particular locations. These are the junctions where the large-angle boundaries of four grains meet, the lines along which three grains are joined, and lastly the surfaces separating two neighbouring grains. It is evident that these different sites do not have the same physical properties and therefore do not behave in the same manner as far as nucleation is concerned. Their specific activity seems to decrease from the quadruple points to the limits between two grains. If we suppose, for simplicity, that only one type of site exists, we will observe the characteristic nucleation rate for this latter.

Let n be the number of sites where a nucleus is formed after time t, or an active site.

For an increase in time dt, we will have dn new active sites. If I is the rate of nucleation, then:

$$I = dn/dt.$$

In the general case, each type of site will have its characteristic nucleation rate, say:

$$I_1, I_2 \quad \text{and} \quad I_3.$$

If these three rates are very different one from the other, the three modes of reaction which they initiate, will be observed in succession.

In many cases these rates do not greatly differ, and the overall nucleation results above all from that of the grain boundaries, since its probability of detection is greater than that of the other two types of site.

Immediately after their formation, the grains develop by displacement of the interface in the lattice.

Let G be the rate of displacement of this interface. We can express the fraction transformed as a function of time if we know $I(t)$ and $G(t)$.

Let $Y(t)$ be the transformation rate, A the area of the interfaces limiting the whole set of nuclei formed at a given time t; we can put:

$$dY/dt = \int_c G dA,$$

if the integration is carried out over all the crystals.

When G is not constant, the solution of the problem is

difficult. It is, in fact, necessary to know a mathematical expression for G, which is not always possible.

This is the case of those alloys in which the discontinuous reaction takes place at the same time as homogeneous precipitation in advance of the front of the discontinuous reaction. The precipitate becomes larger with time, corresponding to a fall in the driving force of the discontinuous reaction. Thus we explain the observed slowing down with time.

If G is constant this is the case for many systems: iron–carbon, Fe–Zn, Au–Ni etc. we have:

$$\mathrm{d}Y/\mathrm{d}t = G \int_c \mathrm{d}A = GA_f .$$

A_f represents the area of the transformed fraction and must be expressed as a function of Y and of t.

When the nodules touch each other, the parts in contact no longer grow, and due to this the active surface diminishes. In order to allow for this effect, we calculate the fraction transformed Y_x which would result from the growth of the nuclei if they were to develop without being impeded by their neighbours. We then have an imagined area and the relationship:

$$\mathrm{d}Y_x/\mathrm{d}t = GA_x .$$

Now $A_f = (1 - Y)A_x$, since A_f is the part of A_x which is outside the volume of the transformed nodules.

Hence:

$$\mathrm{d}Y/\mathrm{d}Y_x = 1 - Y \qquad Y = 1 - \exp(-Y_x).$$

If we know $Y_x(t)$, we deduce Y from this.

The volume v transformed at time t is, in the case of three-dimensional growth:

$$v = fG^3(t - \tau)^3,$$

where f is a form factor.

To find Y_x at time t, we must consider the number of nuclei dn and their volume v, since we have:

$$Y_x = \int_0^t v\,\mathrm{d}n,$$

and since $\mathrm{d}n = I\mathrm{d}\tau$:

$$Y_x = \int_0^t vI\mathrm{d}\tau = \int_0^t fG^3I(t - \tau)^3\,\mathrm{d}\tau.$$

If we know $I(t)$, we can calculate Y_x, and thus Y. If I is constant, we easily find:

$$Y_x = \tfrac{1}{4}fG^3It^4,$$

and

$$\boxed{Y = 1 - \exp(-\tfrac{1}{4}fG^3t^4I).}$$

If the nodules are discs which grow in two dimensions

only, we have:

$$\boxed{Y = 1 - \exp(-\tfrac{1}{3} f \delta G^2 I t^3),}$$

where δ is the thickness of the discs, taken as constant.

Finally, if growth takes place in one direction only:

$$\boxed{Y = 1 - \exp(-\tfrac{1}{2} f \delta^2 G I t^2).}$$

The general expression valid for all cases is:

$$\boxed{Y = 1 - \exp(-kt^n).}$$

The value of n is determined by the type of growth, in one, two or three dimensions, for a constant rate of nucleation.

More sophisticated models of nucleation have been proposed, but they all lead to an expression of the same form.

An interesting case is that where the rate of nucleation is so high that the sites are immediately saturated. Let N be the number of sites.

For two-dimensional nodules:

$$v = kG^3t^3 \qquad Y_x = kG^3t^3N.$$

We reach the general formula:

$$Y = 1 - \exp(-kt^n).$$

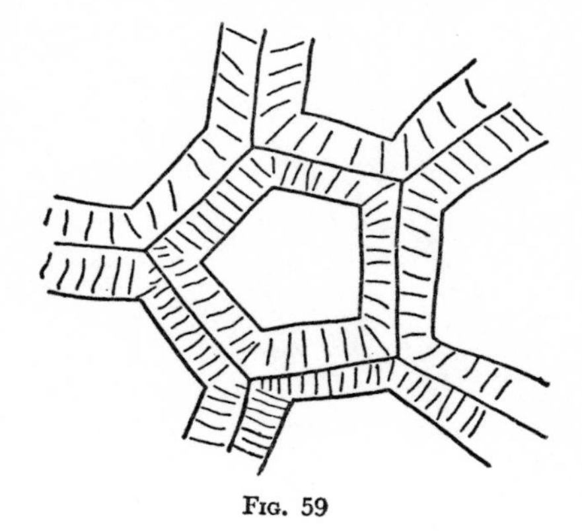

Fig. 59

For quadruple points: $n=3$.

For triple lines: $n=2$.

For surfaces of grain boundaries: $n=1$.

Experimental verification of these theories necessitates measurement of the reaction rate. By adopting a nucleation model and a rate of growth, we can determine the coefficient n and from it deduce the nature of the preferential nucleation sites and the mode of growth. Thus, when the grain boundaries form the active sites, the nodules are numerous and touch each other immediately. The result is the formation of a plane growth front as shown schematically in Figure 59. This is a one-dimensional growth. In this example we find $n=1$. For this to be the case, after measuring Y, we plot $\log[\log(1-Y)]$ against $\log t$. Since

$$\log[\log(1-y)] = A + n \log t,$$

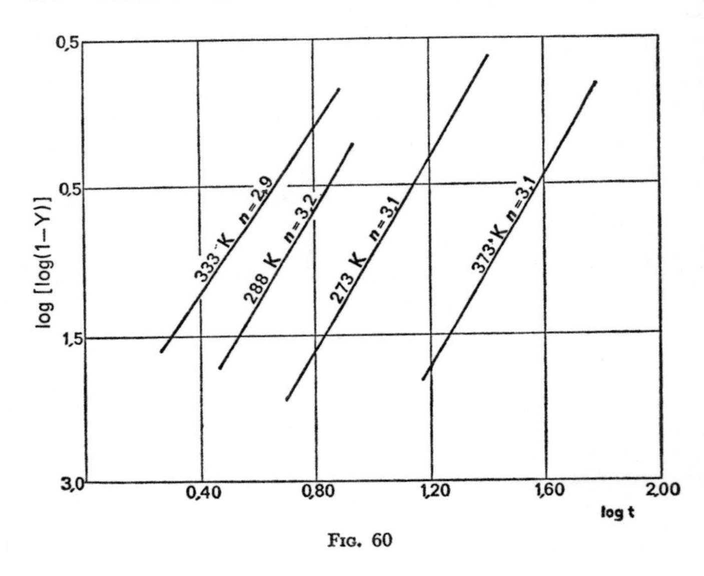

FIG. 60

we must obtain a straight line.

Figure 60 shows the result for an alloy of iron and carbon. The transformation rate was obtained here by magnetic measurements.

We must remark that this method of representation favours the theoretical model, since it always leads to a curve having a straight part of more of less extent. In most cases, however, the theory is verified only for $Y<0.5$.

Since G is the rate of displacement of the grain boundaries, it corresponds to a thermally activated process, and we can put:

$$G = G_0 \exp(-Q/RT).$$

The heat of activation deduced from this expression is frequently smaller than the heat of activation for diffusion in volume, but it is difficult to compare it with the heat of activation of intergranular diffusion, since in most cases this latter is not known.

This type of decomposition is interesting from a practical point of view, in particular where steels are concerned. It would be desirable to develop atomistic theories leading to the design of experiments both better controlled and more delicate than those of recent years.

Lastly, note that these theories apply to the case, where the reaction takes place without displacement of the grain boundaries. Then the progress of the transformation front is certainly facilitated by the existence of stresses at the extremities of the dendrites. The activation energy is smaller than the activation energy for homogeneous precipitation.

CHAPTER 14

MARTENSITIC TRANSFORMATION OF ALLOYS

For centuries, steels have owed their technological importance to the existence of the martensitic transformation. It has been known for a very long time that carburised iron, heated and then suddenly quenched, has an exceptional hardness. Steels with a carbon content below 1% by weight are in the austenitic state when the temperature is in the region of 950° C, for example. If cooling from this temperature is slow, these alloys decompose according to the eutectoid reaction which we mentioned above. If, on the other hand, cooling is very rapid as with quenching by water, there is no time for the carbon to diffuse, and the austenitic structure remains as far as a certain temperature at which it becomes entirely unstable. It then changes into martensite, with very special conditions of production and structure.

This same phenomenon of immediate transformation has been observed in certain pure metals: cobalt, titanium, and lithium, but more frequently in alloys showing a eutectoid reaction such as titanium–manganese, copper–tin, copper–aluminium, uranium–chromium and uranium–platinum.

14.1. General Characteristics of Transformation

The first characteristic of transformation is its rapidity. It takes place over a time of the order of 10^{-4}sec at temperatures between 200° C and absolute zero. It is therefore impossible for it to be a diffusion phenomenon since, at low temperatures, long-distance movements are not possible in so short a time.

A martensitic transformation is defined as a reaction during which the product of the transformation is obtained in a region of the mother crystal by the co-ordinated motion of a large number of atoms, without change in composition. This type of transformation is called a 'military transformation' by Christian, in contrast to the civil transformation which is represented by chemical diffusion where each atom behaves as an independent individual and moves apparently at random. During the martensitic transformation the atoms move only through a fraction of an interatomic distance, and they retain the same neighbours. It is only their relative positions that differ. In particular, if the lattice is ordered, the product of the transformation is itself ordered.

The original material and the product have the same number of atoms, but in most cases there is a change in volume structure and a small change in volume. As a result, if one surface of the original metal is carefully polished, this surface exhibits a set of contours due to the local deformations caused by the transformation. Advantage is often taken of this particular feature to distinguish the martensitic transformation.

14.2. Crystallography of Transformation

The co-ordinated movement of the atoms brings with it a correspondence between the original and the product lattices. Suppose that the elementary cells of the two phases contain one atom only. We can mark the atoms of the material before transformation and make them correspond one by one to those of the product, as shown in the diagram in Figure 61.

The movements of the atoms are such that planes and straight lines of the sites remain planes and straight lines. Only the angles are altered. This is what is known as a *homogeneous deformation.* In particular, shearing is a homogeneous deformation, and it has a corresponding

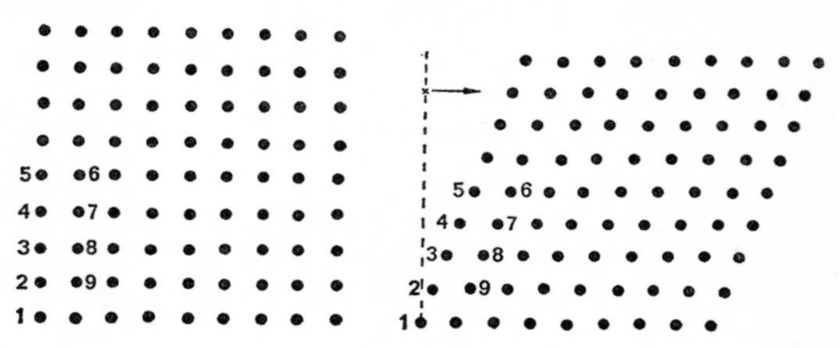

Fig. 61

invariant plane parallel to the shear plane. In this plane atoms retain the arrangement that they had in the original material.

If the elementary cells contain several atoms, it can happen that each atom is displaced differently from the others. In this case there is a superposition of several homogeneous deformations.

In all cases, the change in position of the atoms must show itself as a change in shape of the crystal. Thus, when a small volume is transformed near the surface, the result is a local deformation of the external surface, as shown in Figure 62. The surface is locally at an angle to the initial surface. Measurements of such changes in macroscopic form have been made so as to compare them with the result of the theoretical homogeneous deformation which

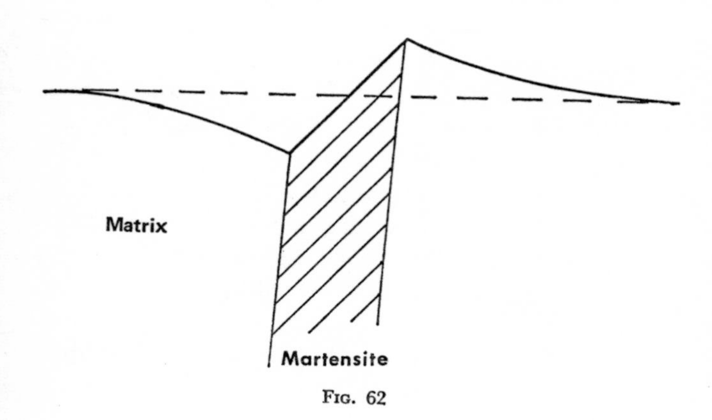

FIG. 62

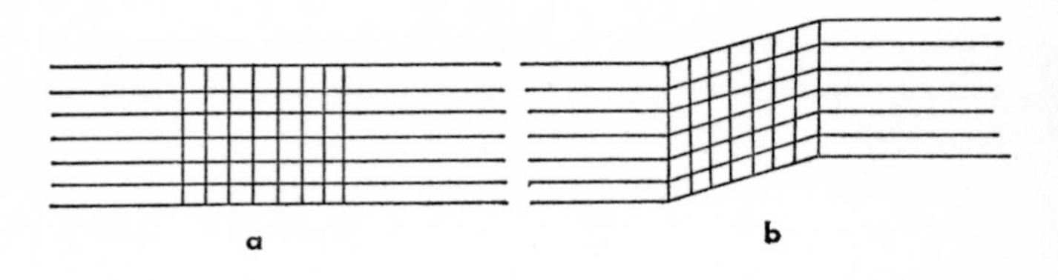

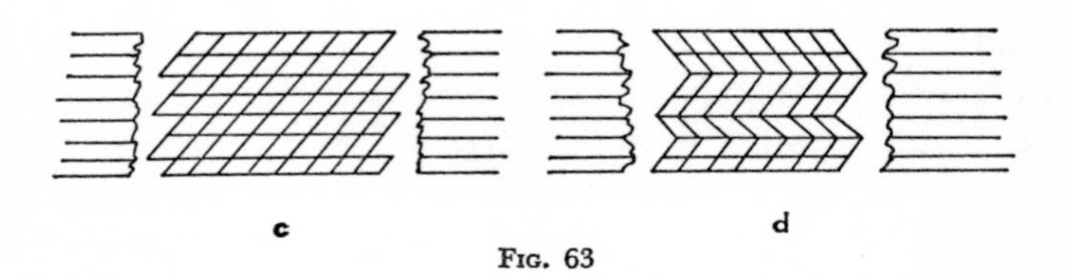

FIG. 63

would have affected the same part of the crystal in the same direction. It is found that supplementary invisible deformations must be introduced. A pure shear is not, in general, sufficient to cause the atoms to pass from one structure to the other. A contraction or an expansion normal to the plane of shear must be added. Further, shearing is not produced by simple means, since this would lead to a substantial change in the surface, which is not observed. The overall deformation due to shear is almost completely cancelled by plastic deformations inside the product. These deformations are themselves homogeneous, since they are the result of slip or twinning.

Figure 63 shows three possibilities for the transformation of one part of a crystal. It is here supposed that the crystal is isolated and that the transformation takes place along one section (Figure 63a).

After homogeneous local shearing, the new structure is bordered by two parts of the untransformed matrix (Figure 63b). These latter are displaced with respect to each other. If the crystal forms part of a polycrystalline assembly, this mechanism cannot operate without considerable deformation of other grains or without formattion of cavities. In fact the deformation of transformation can be approximately compensated for by slip (Figure 63c) or by twinning (Figure 63d). It is quite clear that, if the interface can be perfectly coherent in Figure 63b, the situation is quite different in Figures 63c and d.

We note that in the three cases illustrated, we can consider that the plane of the interface of the matrix has not altered after transformation. It is the plane of shear or invariant plane of the homogeneous deformation.

These three mechanisms lead to orientation relationships between the lattices of the original material and of the product of the transformation.

The following table gives the invariant planes and the orientation relationships observed for different alloys of iron.

Note that slip or twinning does not change the relative orientation of the lattices, hence the theory will be able to predict crystallographic correspondences without being concerned with external rearrangements of the part transformed.

14.3. Structure of the Interface

Since simple orientation relationships exist between the lattices of the original material and of the product, the interface cannot be entirely incoherent. Interfaces are found to be either completely coherent or partially coherent.

Alloy	Fe–C	Fe–Ni	Stainless steel 18-8
Transformation structures	f.c.c. ↓ tet.c	f.c.c. ↓ c.c	f.c.c. ↓ c.c
Orientation relationships	$(111)_A//(110)_M$ $[110]_A//[111]_M$	$(111)_A//(110)_M$ $[211]_A//[110]_M$	$(111)_A//(110)_M$ $[110]_A//[111]_M$
Orientation of discs in the matrix	{225}	{259}	{111}

The transformation which causes, for example, the change from face-centred cubic structure to the hexagonal close packed structure, as in the case of cobalt and Fe-Mn alloys, leads to a completely coherent interface. The invariant plane is in fact one of the {111} planes of the face-centred cubic structure or the base plane of the hexagonal product. These planes have identical atomic arrangements, since their structures differ only in the method of successive addition of layers. By definition the interface does not contain dislocations or discontinuities, inside one and the same crystal of the matrix.

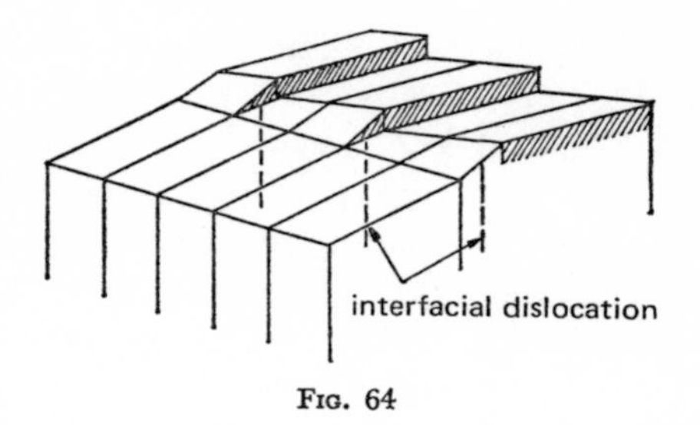

FIG. 64

On the other hand, in connection with the other types of transformation, we must imagine models describing the connection of the two lattices whilst still allowing growth of the product by rapid displacement of the interface parallel to itself. One model which is preferred by many theorists to account for the martensitic transformation is that of screw dislocations in the plane of the interface.

These dislocations belong to both lattices, and as they slide in the planes of the matrix, they permit the volume transformed to increase without introducting a large amount of energy. Figure 64 shows a diagrammatic arrangement of such interfacial dislocations. It appears that this model is not entirely consistent with the complexity of the phenomenon.

14.4. Morphology of the Products of the Martensitic Transformation

The most typical form of martensite is that met with in

steels. It is composed of irregular sheets with random orientations, in the form of small discs which are easily recognised by microscopic observation. This martensite is known as *acicular*. It appears in high-carbon steels or in iron-nickel alloys. It is also met with in some alloys of other metals. It appears characteristic of alloys with a low transformation point, say in the region of room temperature.

Other alloys are transformed by displacement of a plane interface parallel to itself. This takes place in the case of the transformation face-centred cubic to hexagonal close packed, cobalt alloys or manganese steels. We then observe the formation of narrow parallel bands grouped into four families parallel to the $\{111\}$ planes of the matrix.

Finally, the martensite known as massive is met with in many systems. First observed in copper-aluminium and copper-zinc alloys, it has been recently found in low-carbon steels where the transformation point is fairly high (between 200° and 500° C. The product of the transformation appears on microphotographs formed from irregular grains with straight joints interrupted by transcurrent discontinuities. The transformation gives rise to a change in volume which is seen as an unevenness on a face which was previously polished. The simpler form of the grains in massive martensite, compared with acicular martensite, comes from the higher transformation temperature. Since the stresses of the transformation can be more easily relieved for massive martensite, the product of the transformation has more degrees of freedom insofar as its form is concerned, than is the case for the discs of acicular martensite. It appears in fact that the disc-like shape minimises the energy of deformation as it does for precipitation.

Observation by means of the electron microscope of the different types of martensite has given information about their fine structure.

The discs of acicular martensite comprise a central part finely twinned, and lateral parts where the density of dislocations is very high.

Massive martensite is formed from elongated grains where the density of dislocation is very high.

Lastly, martensite of cobalt alloys is formed from plane sheets with a fairly small number of dislocations but many faults between layers. These latter correspond to laminae of hexagonal compact structure resulting from a crystallographic change in phase between two adjacent grains.

14.5. Thermodynamics of the Reaction

The martensitic transformation is not accompanied by any change in concentration. It is therefore comparable to an allotropic transformation. We can represent the free energy of the alloy as a function of the temperature (Figure 65) for the structures corresponding respectively to the matrix and to the product. This graph assumes a linear relation between G and T. The two lines intersect at a

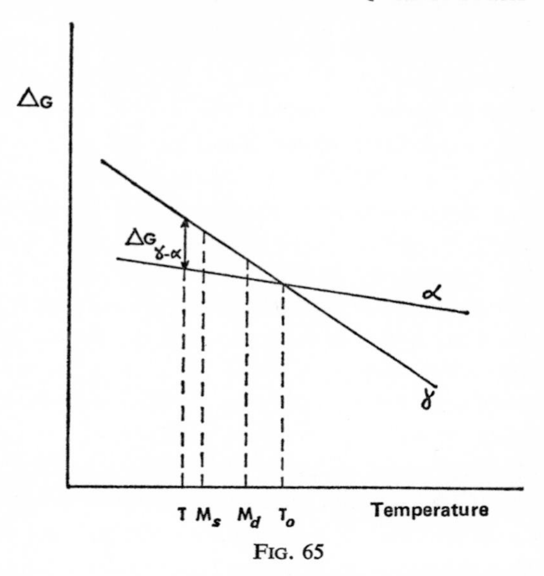

FIG. 65

point T_0 and we see that above T_0 the matrix is stable, and below this temperature it is unstable.

The change in free energy $\Delta G_{\alpha-\gamma}$, when the crystal passes from one structure to the other is negative, zero, or positive. The driving force of the reaction is equal to $\Delta G_{\alpha-\gamma}$.

The transformation is in fact accompanied, as we have seen, by deformations which create stresss in the matrix and in the product. The result of this is an internal energy of deformation which absorbs part of the energy liberated by the transformation. As with precipitation, the nucleus of martensite is limited by an interface to which there also corresponds a postive free energy.

The surface and deformation energies vary only slightly with temperature, whilst the change in free energy in absolute value, increases as the temperature falls. It is therefore necessary to reach a temperature below T_0 for the transformation to take place. This temperature is the martensitic transformation temperature M_s.

The difference $T_0 - M_s$ varies with the composition of the alloys and its value depends apparently both on the properties of the matrix and on those of the product of the transformation. The temperature M_s can correspond during cooling to the instant when resistance to the transformation vanishes, or again to the instant when the activation energy becomes sufficiently small for the reaction to take place.

The transformation can always be obtained above M_s by application of an external stress which neutralises the internal stress. We then reach a definition of a transformation temperature induced by the deformation. This is the point M_d between M_s and T_0. Above M_d the matrix is deformed without being transformed. It can, however, be transformed in the course of cooling, if M_d and M_s are above room temperature.

Theoretical calculation of the position of the points M_s and T_0 has been attempted, but it remains phenomenological, without introducing the local changes of structure which certainly play a major part.

14.6. Martensitic Nucleation and Kinetics of Growth

As for certain aspects of precipitation, the formation of nuclei of martensite can be explained by Becker's theory. At a given temperature, the formation of a nucleus is only accompanied by a negative change in free energy if the change in free energy of transformation (passage from one structure to another) is greater than the energy of deformation and the interfacial energy. There is therefore a critical size and energy of activation which are calculated in the same manner.

The rate of nucleation is of form:

$$I = K\nu \exp(-w/kT),$$

where w is the activation energy to be supplied to form a nucleus of critical size; ν is the frequency of vibration of the atoms, connected with the temperature.

The activation energy falls as the temperature becomes lower. It seems that the phenomenon is athermal in most cases. However, as far as Fe-Ni-Mn-C alloys are concerned the transformation is isothermal and the rate of transformation depends on the temperature.

At high temperatures, nucleation is difficult and the reaction is slow.

At low temperatures, nucleation is immediate but the thermal agitation is reduced, and the reaction takes place very slowly. There is therefore an intermediate temperature at which the rate is a maximum. Figure 66 shows the graph of temperature against transformation rate as a function of time for alloys of Fe–Ni–Mn–C.

The athermal transformation is characterised by the fact that, at a given temperature below M_s, there corresponds a definite rate of transformation. There must be a further fall in temperature for a supplementary volume to be

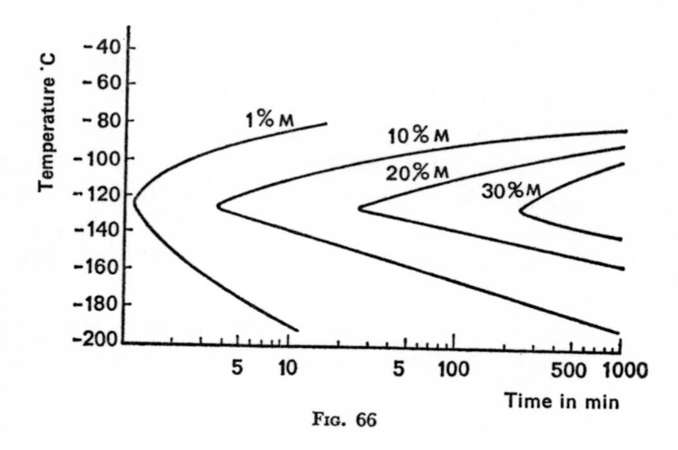

Fig. 66

formed. The reaction is immediate and can take an explosive nature. The rate of displacement of interfaces between matrix and product is close to the speed fo sound. It appears to be practically independent of temperature. The kinetic depends therefore only on the nucleation. The transformation is stopped by the grain boundaries or by the defects accumulated by the correlative deformation.

Thus, the kinetics of the athermal martensitic transformation are directly linked with the rate of cooling of the alloy.

The isothermal martensitic transformation is preceeded by the formation of athermal martensite. If cooling is stopped, the reaction continues whilst slowing down. The region of temperatures where the transformation is observable is found to be below room temperature.

The structure of isothermal martensite is similar to that of athermal martensite. It seems that the increase in the volume transformed results most frequently from the formation of new nuclei and not from the growth of nuclei already formed. Nucleation is therefore in different places and at different times.

Secondary effects have been reported: the autocatalytic effect and thermal stabilisation.

The autocatalytic effect corresponds to a nucleation induced by stresses of transformation round a nucleus. It results in an explosive transformation.

Thermal stabilisation is observed when a sample is maintained at a temperature slightly above the temperature M_s or even at the transformation temperature. In the first case the starting temperature of the transformation is below M_s. In the second case a lower temperature must be reached for the transformation to recommence. This effect is observed in iron–carbon or iron–nitrogen alloys. It appears that the mobility of the interstitial elements allows a rearrangement which blocks the transformation. It is then necessary to increase the driving force in order to restart the reaction.

14.7. Properties of Martensite

We can consider two types of martensite: that of carbon or nitrogen steel and that of substitution alloys.

The martensite of carbon steels is of considerable technological importance. This is the structure which gives steels their greatest mechanical resistance. In addition, its formation and properties can easily be controlled by adjustment of the concentration of the carbon and other elements added, for example, to facilitate quenching, or by subjecting it to annealing treatments which lead to a lower mechanical resistance than in the quenched state but which allow it to recover a certain degree of ductillity.

The origin of the hardening seems to be linked to the solubility of carbon. Solubility of carbon is, in fact, very large in austenite, of the order of 1% by weight, whilst it is very small in ferrite, less than 0.001% by weight. The carbon is trapped when quenching takes place; its presence causes deformation of the ferrite lattice which it renders tetragonal (Figures 67a, b). The mechanism of hardening of martensite by carbon is, however, still not very well understood. Some part of the mechanical resistance must be attributed to the presence of numerous dislocations and twins formed during transformation.

On the other hand, in the other type of martensite of low-carbon or low-nitrogen alloys, the martensitic trans-

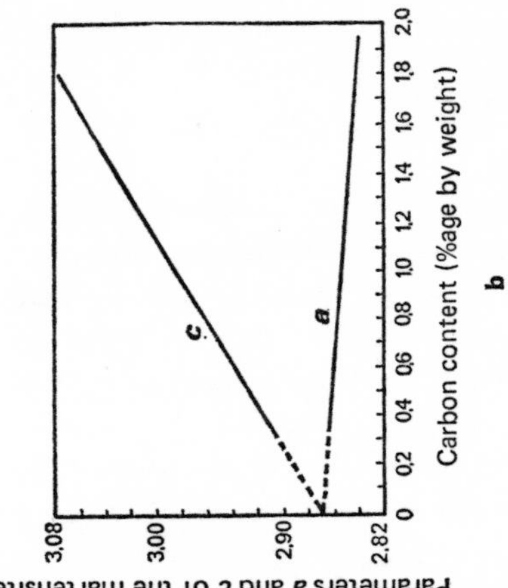

c
a
a
A
B
C

Possible positioning for iron atoms

Probable positions of carbon atoms

a

FIG. 67

formation also increases the hardness. Thus, iron–nickel alloys have a hardness of 100 HV 30 in the austenitic state, whilst their hardness is 250 HV 30 in the martensitic state. In this case the increase in mechanical resistance is due to the division into small crystals and the presence of dislocations caused by transformation.

BIBLIOGRAPHY

J. Bernard, A. Michel, J. Philibert, and J. Talbot, *Métallurgie générale*, Masson, 1969.

J. Burke, *The Kinetics of Phase Transformations in Metals*, Pergamon, 1965.

B. Chalmers and R. Kings, *Progress in Metals Physics*, V, Pergamon, 1954.

A. H. Cottrell, *Métallurgie structurale théorique*, Dunod, 1955.

G. Emschwiller, *Chimie physique*, Presses Universitaires de France, 1951.

M. E. Fine, *Phase Transformations in Condensed Systems*, Macmillan, 1964.

E. A. Guggenheim, *Mixtures*, Clarendon Press, 1952.

A. Prince, *Alloy Phase Equilibria*, Elsevier, 1966.

P. Souchay, *Thermodynamique chimique*, Dunod, 1955.

R. A. Swalin, *Thermodynamics of Solids*, Wiley, 1961.